Edizioni PensareDiverso
Cenacolo Jung Pauli

CENACOLO
JUNG PAULI

Bruno Del Medico
Illustrated by Elisabetta Del Medico

All the Colors of Quantum Entanglement

From the myth of Plato's cave, to the synchronicity of Carl Jung, to the holographic universe of David Bohm. Quantum physics rejects materialism and reveals the spiritual component of the universe.

Summary

Introduction

The book is divided into three parts. In the first part (The intuitions) the author deals with the most relevant hypotheses on the illusory reality of the perceptible world. The existence of a level of consciousness that transcends matter has been envisaged by the great thinkers. We find this idea in Plato's Myth of the Cave, in Berkeley's Immaterialistic Theory, in the Psychology of Form. The most authoritative source lies in the works on the collective unconscious and on the theory of synchronicity by Carl Jung.

In the second part (Confirmations) the author describes in an elementary but detailed way the path of quantum physics, from Thomas Young's double slit experiment to the phenomena of the superposition of states and quantum correlation. Through these privileged keys it is possible to understand quantum entanglement. In the third part (Perspectives) the author describes the theories developed by David Bohm on the "quantum potential" on the "implicated universe". and on the holographic vision of the cosmos. Everything is explained with absolute simplicity, without the use of mathematical formulas and with the help of many illustrations.

Humanity, from its very beginnings, wanted to investigate the origin and composition of things, to discover their functioning and their intimate purpose.

The universally used method is to break down objects into smaller and smaller parts, then analyzing them with every possible technique, from visual investigation to chemical reactions. This still happens today. For example, if a scientist wants to discover the chemical and physical structure of a cube of granite, he will break it into smaller and smaller pieces until it is divided into individual atoms.

However, if the scientist himself wants to investigate the individual particles that make up the atom, he receives an incredible surprise. The granite cube behaves like an ice cube would. The scientist sees the matter that becomes fog, evaporates, disappears between his fingers. Solid matter becomes energy that vibrates.

The single particles are transformed into fluctuating waves without any solid corporeality.

At the subatomic level, matter is no longer solid matter, it becomes something different.

Elementary particles deceive us. They look like solid specks if someone observes them, but they behave like vibrating waves when they are not observed.

Atoms practically only contain vacuum.

On the surface, we believe we can touch, weigh, manipulate and measure matter. But, in its most intimate composition, matter becomes a ripple of emptiness, energy, information, wave or vibration. What seems to us solid material, in its most intimate essence is no longer solid material.

At this point, it is clear that we can no longer speak of a single reality. Depending on the levels of observation, from the extremely small to the infinitely large, there are many realities, all different but all absolutely true.

Or, perhaps, there are many aspects of a higher reality, still unknown. All philosophies and religions have always hypothesized a "zone of the spirit" transcending matter; no one, however, has ever been able to provide proof of its existence. Today quantum physics is opening a huge window on horizons that, until the last century, we could not have imagined. The confirmations come from the experiments carried out successfully, especially those relating to the phenomenon of quantum entanglement.

Today we know that there is a level of reality that is no longer subject to the constraints of Newtonian physics. The physics of matter is no longer sufficient to describe the universe.

Quantum physics demonstrates the existence of a level in which energy and information take over matter. This is the so-called "non-local" level. We could define it a psychic or spiritual level. At this level, a universal Intelligence interacts with humanity. The paths of communication with the intelligent universe pass through the collective unconscious that was theorized by Carl Jung.

Jungian synchronicities guide us in a cultural evolutionary project. It is a project that we are beginning to become aware of.

First part. Insights

Unexplained phenomena

According to some renowned aeronautical texts,
the hornet cannot fly due to the shape
and weight of its body
in relation to the surface of its wings.
But the hornet does not know,
and therefore continues to fly.
(Igor Ivanovich Sikorsky, Russian aviation pioneer)

The Guinness Book of Records contains many oddities. Among these unusual news there is a really special one. This is the ranking relative to pairs of twins born at different times. Guiness signals which couples are born at a greater distance of time.

Contrary to what one might imagine, this interval does not consist of hours or days, but even months. For example, in 2000 Sandra Beveridge of West Lothian, a Scottish county, gave birth to twins 28 days apart. Peggy Lynn, a lady from Pennsylvania, did even better. In 1995 Peggy gave birth to twins born at 84 days apart. However, the current record belongs to Maria Jones-Elliot who in 2013, in Ireland, gave birth to Amy and only after 87 days gave birth to the twin Katie.

The doctors specify that these are very rare cases. These cases occur only when the twins are raised in separate placental sacs.

Quantum-related twins

There is an interesting aspect, which convinced me to mention the primacy of delayed twin births. It is the fact that the book "The Guinness Book of Recods" was born from the pen of two authors. Their names were Ross and Norris McWhirter, and they too were twins. These two twins, in addition to being the authors of the book, were the protagonists of an "impossible" event.

Currently Guinness is the third best-selling book in the world, after the Bible and the Koran. This book represents the practical development of an idea born in a completely random way, during a hunt that took place on May 4th 1951.

One of the participants in the hunt was Sir Hugh Beaver, managing director of the Guinness breweries in Dublin.

He could not hit some golden plovers. The birds were faster than his rifle and fled with great speed as soon as they felt the presence of the hunter.

This inconvenience gave rise to a debate between Sir Beaver and the other hunters, to establish which was the fastest bird.

In the following days Sir Beaver, who evidently was not very busy with the role of administrator, continued to brood over the problem. In the end he came to the conclusion that many, like himself, would be interested in knowing not only which was the fastest bird, but every other type of excellence in different performances. He therefore decided to create a book where everyone could access this type of information.

The content of the book was to be a collection of records in the various activities. As a result Sir Beaver asked for help from two athletics specialists, the brothers Ross and Norris McWhirter and commissioned them to write the book.

Figure 1 - The twins Norris (left) and Ross McWhirter (right) authors of the first edition of "The Guinness Book of Recods". The two were protagonists of an incredible case of extra-sensory communication

The book was published for the first time on 27 August 1955 with the signature of the two brothers. The title was "The Guinness Book of Recods". The book immediately had that resounding success that continues today. (Fig.1).

Obviously, the authors became well known, so the events of their lives are very well documented and witnessed.

The twins Ross and Norris McWhirter

Ross and Norris were born on 12 August 1925 by William McWhirter, editor of Sunda Pictorial, and by Margaret Williamson. In 1950 they both became sports journalists. A friend of theirs, the runner Christopher Chataway, was employed at the Guinness brewery.

When Christopher learned that administrator Hugh Beaver was looking for competent people to make a book on records, he recommended the twins. The two immediately received the assignment.

In the 1960s Ross became a political activist for the Conservative Party. Regardless of difficult times, he supported many restrictive laws promulgated by Britain against the Irish population.

He even promoted a reward of £ 50,000 to anyone who had provided useful information for the arrest of an IRA (Irish Republican Army) cell. At that time the IRA was committing many attacks in London. The IRA considered Ross McWhirter as a direct enemy.

On November 27, 1975, two members of the IRA ambushed the journalist. They waited for him as he returned home, at 6.45 pm, and fired two shots at him with a .357 revolver.

At the same time that Ross was killed, twin Norris was about 30 miles away and was talking to some people. Suddenly, at 6.45 pm, he grabbed his head in his hands, as if it were bursting. To the astonishment of those present, he almost lost his senses.

The people present witnessed the fact. The testimony is documented by the scholar Guy Lyon Playfair in his book "Twin Telepathy":

> "At the beginning of the evening of November 27, 1975, the writer and television personality Ross McWhirter was shot in the head and chest by two armed men while he was standing at the door of his house in north London. He was hospitalized, but was pronounced dead shortly after arrival. I got an account from an eyewitness who was with Norris McWhirter. Norris had a dramatic reaction, almost as if an invisible bullet had hit him. "

This is not the only case of a twin who, being at a great distance, reacts to events involving the other twin.

Viscountess Thelma Furness

Thelma Morgan, born in Switzerland on August 23, 1904, was the daughter of Harry Hays Morgan, an American diplomat who served as US consul in Buenos Aires and Brussels. She was brought into the world by Laura Delphine Kilpatrick along with her twin sister Gloria. (Fig.2).

Laura Delphine was the nephew of a Union general and, through her maternal grandmother, was a descendant of the Spanish royal house of Navarre.

Therefore, the twin sisters had all the credentials to attend the best society of the time. Thelma became the lover of the Prince of Wales, the future king Edward VIII. Therefore, he preceded Wallis Simpson in the graces of the king. Later, for Wallis's sake, Edward abdicated and became Duke of Windsor.

During his relationship with Prince Edward in 1929, Thelma married Viscount Marmaduke Furness. The marriage did not last long but allowed her to boast life naturally during the title of Viscountess Furness. During the short marriage Thelma gave birth to William Anthony.

Thelma's twin sister, Gloria, married Reginald Claypoole Vanderbilt.

Despite the diverse and eventful lives, the two sisters maintained a strong psychic bond. This bond manifested itself blatantly when

Thelma, pregnant with William, had to face premature birth while at Melton Mowbray, in the United Kingdom.

At that time his sister Gloria was at the other end of the ocean, in New York, so she was completely unaware of the premature birth. However, simultaneously with what was happening to her twin, Gloria experienced severe abdominal pain, just as if she herself was giving birth. Gloria was promptly rescued but it was never possible to establish the origin of her pains. Only later was it possible to link the pains of Gloria to the contemporary childbirth of Thelma. Gloria mysteriously perceived the difficulty that Thelma was going through and participated as if she had been one with her. The union was not only psychic but also physical.

I wanted to mention two well-known cases, on which there are abundant and sure testimonies. These cases confirm that some pairs of twins, by virtue of a mysterious bond established by common birth, can share physical sensations in a way that contradicts every elementary law of physics. This is more evident if we consider the distance of the twins when the facts occur.

Figure 2 - Thelma Morgan, Viscountess Furness (right) with her twin sister Gloria. While Thelma, in the United Kingdom, was suffering from the pain of a premature birth, her sister Gloria, in New York, experienced the same pains.

The notoriety and the testimonies of these cases make it difficult to maintain that these are invented stories. These episodes are so well documented that they are also mentioned on the CICAP website (Italian Committee for the Control of Claims on Pseudo-sciences).

This association has taken on the mission of challenging any claim relating to extrasensory phenomena. However, the CICAP catalogs the cases described as "sensory coincidences" and defines them "decidedly sensational cases".

In reality, the stories of these twins show a great correspondence with some typical aspects of the subatomic particles. Quantum physics has shown that two correlated particles, that is joined by particular conditions (typically, the birth from a common event) acquire the ability to remain "consciously" connected at any distance, as if they were a single particle. We will talk about it in the second part of the book.

Matter, only matter, nothing but matter

"Materialists and fools never have doubts."
(Gilbert Keith Chesterton)

According to current science, there are things that can be, and others that cannot be.

The first are all those things that can be weighed, measured and reproduced in the laboratory. Among these things we also understand the related effects, such as magnetism and gravity.

All other instances, starting with extrasensory phenomena, are the result of illusion or superstition, and their supporters are cheaters or, at best, suffer from mental disorders and should be subjected to medical treatment. To support this thesis, absurd explanations are sought that border on the ridiculous. In any case, the existence of these phenomena is denied in advance.

It must be said that the scientific forum rests this attitude on apparently solid bases, that is, on the known physical laws. These laws have been affirmed and consolidated for some centuries. These are the laws that we ourselves can verify by observing the world around us with our deceptive senses. I will try to summarize briefly the materialistic theories.

Monism and materialistic reductionism

The concept of monism is very ancient, dating back to the 5th century BC, specifically to the philosophy of Parmenides. The real term of monism was used in 1734 by the German philosopher Christian Wolff who, in his work "Rational psychology", defines "monists" as those who "admit a single kind of substance".

Monism is distinguished from dualism or pluralism, according to which the whole is attributable to two or more kinds of substances. Some typical dualisms are mind / matter, or mind / body.

Materialist science reduces all reality to a single kind of substance, matter. At the same time materialism excludes any psychic or spiritual presence. Only everything that is physical is real. The mind, the thought, the consciousness are only the residue of chemical processes that take place in the brain.

Incidentally, we specify that there is not only materialistic monism; for example, there is also a monism called an idealist. This is called "mentalism", and it states exactly the opposite: only the mental aspect is real.

Both of these positions are not corrected in light of recent discoveries in quantum physics, as we will see later in the book. In fact, both deny that a universe can exist in which matter and psyche (or spirit) collaborate harmoniously in a great creative project.

Determinism and causality

According to these principles the universe is a machine that works like a clock. Everything is established, everything is determined. Each event, each change of an object is functional to its position, its initial speed and the material forces present.

A seed absorbs nutrients from the soil and produces a seedling. The seed pushes the shoot upwards with a certain force. If the force is sufficient, the shoot breaks the crust of the soil and emerges, otherwise it dies underground. As soon as a sprout emerges, the force coming from the roots pushes it upwards.

At that point he can compete with the surrounding plants in absorbing sunlight and rain. Everything works like a clock, the plant grows by a series of mechanisms, of which the plant is not absolutely aware.

If I throw a stone, this will take a path determined by its weight and the force with which I threw it. If the stone hits an object, it gives the affected object a boost proportional to the energy possessed. It will never happen that a stone throws itself from itself, or that an object moves, if it is not solicited by another object.

Therefore, every action is caused by another action. Every action inherits strength from the action that caused it and distributes strength to the subsequent actions.

This is very useful to current science because it allows us to determine the position of each object in the universe at any time, and to predict the behavior of machines and industrial equipment. The machines, if they receive the appropriate energies, can perform precise functions up to the thousandth of a millimeter to obtain the desired results.

René Descartes split reality into mind-matter dualism, thus anticipating the concept of determinism, which excludes the need for a mind. The concept of exclusion of the mind in favor of the uniqueness of matter was then developed by Isaac Newton and Pierre Laplace, in his book on celestial mechanics. (Mécanique céleste, 1805).

This was the first book not to mention God as the cause of creation. The emperor Napoleon himself was astonished and asked Laplace: "Why did you not mention God in your work?"

Laplace replied "It's a hypothesis I didn't need".

Causality absolutely denies the possibility of a creative project in which man participates. Human intelligence is a useless by-product of matter. Everything works very well without the need for any kind of intelligence.

The continuity and the arrow of time

The statement of science is that every event is caused by another event. In turn, every event causes others. This implies the concept that all the changes and all the dynamics of the universe follow each other continuously, without any pause, in a carousel that began with the Big Bang and will never end. Moreover, all this continuous evolution takes place in only one direction, according to the arrow of time that proceeds only forward.

We call "time arrow" a phenomenon such that a physical system can evolve from an initial state to a later state, in a given time, but it can never return to the initial state.

Reality as we perceive it is called the reality of the macroscopic level. At this level we are subject to the law of entropy. This law says that disorder can only increase with time. So, if you break a glass, you will certainly be able to break each piece even further.

However, forget that the different pieces are reconnected to form the healthy glass again.

Curiously, we can see that this is true only at the macroscopic level. Instead, at the microscopic level, almost all physical processes are symmetrical. That is, the equations that describe them show that they would work very well both forward and back in time.

Strangely, according to the principle of causality, science is able to predict exactly the behavior of the universe that surrounds us.

However, at the same time science denies that we can know in advance what will happen next in our personal lives. Thus, science rejects the existence of the phenomena of premonition. That is, science claims to predict the future for the universe, but denies the possibility of predicting the future for the individual.

The concept of "non-locality"

All phenomena and their effects are only localized, that is, they occur in a precise portion of space. Furthermore, they propagate with a speed and time that can be measured accurately.

An action at a distance between two objects without the mediation of a "physical force" that connects them is impossible. It is necessary that "something" unites the two objects through the space that separates them in a certain time. This force fades according to the length of the path.

Material objects have a limit speed of about 300,000 kilometers per second. Nothing can overcome this speed. This determines, for example, that communications between space satellites and the planet's surface can be delayed even a lot.

If an eruption occurs on the surface of the Sun we can see it only after eight minutes. This is the time taken by the light of the explosion to travel to the Earth. So, if you happen to look at the Sun, know that that's the Sun eight minutes ago. Meanwhile the sun may have exploded, but we'll only know it after eight minutes.

This feature prevents a certain action taken in one place from having contemporary effects in another place.

So, to give a spatial example, we can assume that if a mad asteroid escapes the Cloud of Oort and points straight to planet Earth, we will have at least a few years to develop defensive strategies.

In this sense, science has enthusiastically accepted Einstein's confirmation, according to which the concept of locality remains valid even in the context of the theory of relativity.

Fortunately, the speed of light limit prevents the asteroid's escape from coinciding with an immediate impact with our planet.

At the macroscopic level, that is in the dimensions of the universe, the known physical laws prevent this from happening. Even our daily life takes place at a macroscopic level. Local laws guarantee that if, within the framework of a family dispute, the spouse throws a plate at us, we have time to move before being hit.

As we will see, on the subatomic level, instead, the "law of nonlocality" applies. At this level the plate would hit us at the same instant it is launched, whatever our distance.

Objectivity.

The material world is absolutely independent of human thought, that is, of consciousness. Consciousness has no weight or mass and therefore cannot make an object move at a distance, as a billiard ball could do. No elaboration of the consciousness can modify the matter or influence it.

Yet, there are situations, at the subatomic level, in which the observer, with his will and therefore with his conscience, determines the reality. With his method of observation the observer can establish the nature of a subatomic object, that is, he can determine whether that object is a wave or a particle. Not only that, the observer can determine the behavior of the object. Thus, at the subatomic level, we are seeing an objectivity to the contrary. We will verify this aspect in the chapter dedicated to the double slit experiment.

Illusory symptoms

Again according to supporters of materialism, all that is not material is an unwanted product of physical matter.

Consciousness, thought, feelings, love or piety are accessory manifestations, "spurious" of matter, and their presence or absence does not affect the performance of real phenomena. The phenomena can exist only if they are physically related to the matter itself. Spirituality and all mental elaborations are waste products of the chemical processes that take place in brain matter.

Yes, however, the Gödel completeness theorem ...

So much scientific security is immediately questioned by science itself. Kurt Friedrich Gödel was a naturalized American mathematician, logician and philosopher, best known for his work on the incompleteness of mathematical theories.

The theorem that bears his name basically says one thing. Any algorithmic system used to test a mathematical principle turns out to be incomplete or contradictory, since it will surely contain a part that cannot be proved.

This happens when the algorithm is sufficiently processed.

Physical laws are also mathematical. Therefore, they are based on mathematical algorithms that inevitably contain a logically unprovable part, which however works.

This means that, when a mathematician elaborates an algorithm, he exits the logical system to capture a mathematical truth that transcends the context in which the logical procedure and the demonstration are valid.

This truth already exists, but cannot be "caged" in the formula. Reality is perceived and used instinctively without being able to make it "systematic".

Therefore, if the laws of physics guide the behavior of matter, it must be recognized that they contain elements that exist a priori, which transcend the material world.

This idea is called "mathematical Platonism" since Plato had already understood that the behavior of the material world is subject to mathematical symbols that do not belong to the logical process but to the archetypal world.

Mind and brain

Is the universe nothing but a jumble of matter, or does it have its own consciousness, as the most ancient cultures have always affirmed?

Is our consciousness a by-product of brain chemistry or does it come from universal consciousness?

The theory of very many cultures is that "universal consciousness" is the source from which all creation flows. Everything is alive and everything is one.

This conception that we can define as mystical has been denied and rejected by the scientific culture of the last centuries. It is a culture that developed nobly in the Enlightenment but then became corrupted in materialism.

Currently, most scientific circles believe that the mind is just a brain creation. man would only be a machine that, as a secondary effect, has acquired the ability to process thoughts and feelings. If we had machines capable of analyzing the brain in depth, we could establish with certainty what chemical reaction generates in us the feeling of having a soul, of loving our mother or of believing in the existence of eternal life.

The hope of materialism is that one day a "brain detergent" can be developed that inhibits the formation of these troublesome chemical reactions, freeing humanity from all superstition.

Thus, science considers the phenomenon of life starting from the bottom, ie from matter as the primary basis. The matter is aggregated into molecules and cells, which evolve and rise on the evolutionary scale under the guidance of a DNA born spontaneously from the case.

The DNA generated by chance randomly generates the brain. In the brain, electrochemical stimuli generate the dispositions that make the body machine work. As a secondary product, the brain also generates dreams, feelings, emotions, creativity and everything we call the mind's activity.

Conversely, all the great spiritual traditions consider the phenomenon of life starting from above, not from below. In the first place there is a universal conscience from which everything emanates, first of all life.

In every form of life the universal consciousness continues to be present, and every life form has its own individual consciousness. Individual consciousness is a fragment of universal consciousness. All creation emanates from universal consciousness, including matter. Matter is also permeated with universal consciousness.

Thanks to this, matter is harmoniously combined according to a project that is external to physical matter.

The myth of the cave

We can forgive a child when he is afraid of the dark. The real tragedy of life is when a man is afraid of light.
(Plato)

How is a chair made?

There has always been a cultural current that, through the various disciplines of thought, from philosophy to spirituality to modern science, has supported the illusory nature of the reality in which we live. Theories of the most diverse genres have been created. In the poem "On nature" Parmenides claims that the multiplicity and changes of the physical world are illusory. As we shall see, Plato in the "Republic" proposes the famous myth of the cave. Later in time George Berkeley will argue that we have a feeling that things actually exist but, in reality, things are just abstractions. According to Berkeley there are no bodily objects, only collections of ideas.

Until a few decades ago these were considered curious and totally imaginative theories. Nowadays quantum physics has shown that these are ideas absolutely endowed with their own dignity. In fact, at our macroscopic level we examine "things" and consider them concrete. However, "things" completely change appearance if we examine them in the infinitely small.

A chair seems to us to be undoubtedly a solid object. The same chair, examined at the level of the atoms and the subatomic particles that compose it, becomes a set of vibrations scattered in an immense void.

All matter, examined in its subatomic composition, has these characteristics. Everything that appears tangible, weighable and measurable is such only in function of what is perceived by our senses. Therefore, the senses show us an illusory and imaginary cosmos. It must be so to allow us to live, to move, to operate in the environment that surrounds us. This does not prevent all things from being a gigantic illusion dimensioned on our physicality. In practice, we live in a dollhouse with all the furniture and furnishings custom built according to our needs and needs.

Outside the house, everything is very different. As the protagonist of a famous Shakespeare tragedy says, "There are more things in heaven and on earth, Horace, than your philosophy imagines." (Hamlet, act I, scene V).

According to some surprising experiments, everything we perceive as "reality", including our own bodies, could be the result of

Figure 3 - Plato (Athens, 428/427 BC - Athens, 348/347 BC), was an ancient Greek philosopher. In "Timaeus" Plato discusses the relationship between the world of nature and the world of ideas..

the interactions that occur between our consciousness and the light, chemical, electrical and those stimuli of any nature perceived by our senses.

In practice we live in a vision of reality that surrounds us completely reworked by our consciousness. We know very little about this reality, how it really is made.

So what is reality really? If we refer to what we can touch, smell, taste or see, we are on a false path. All these "sensations" are only chemical, electrical or other signals. Our senses perceive these signals and send them to the brain. The brain then builds an image according to little-known procedures.

If we look at a tree in front of us, the sight perceives its shape and sends it to the brain. The image of the tree is constructed in the brain. The brain attributes the colors, scents, substances and other characteristics to the shape of the tree. Beyond the mediation of the senses, we do not know how the tree is made, nor what its colors are, nor do we know if, in reality, it is colored

Plato and the search for knowledge

Plato was an ancient Greek philosopher. Together with his teacher Socrates and his pupil Aristotle he laid the foundations of western philosophical thought. (Fig. 3). He was born in Athens to aristocratic parents: his father Ariston boasted among his ancestors the last king of Athens, Codro, while it is said that his mother, Periction, descended from the legislator Solon.

His father gave him the name of Aristocles, his gymnastics teacher. Aristocles himself gave him the nickname Plato, both for the width of his shoulders and to prevent the pupil from having the same name as the teacher.

According to Apollodorus of Athens, Plato was born at the beginning of the 88th Olympiad, on the seventh day of the month of Targellion, i.e. at the end of May of 428 BC. Plato had two brothers, Adimanto and Glaucone, who are mentioned in his work "Repubblica". He also had an older sister, with the curious name of Potone.

Speusippo, son of Potone, praised the intelligence and prodigious memory shown by Plato when he was a child.

Plato participated in three military expeditions, during the Peloponnese war, in Tanagra, Corinth and Delio.

After meeting Socrates, Plato destroyed all his poetic compositions and devoted himself completely to philosophy.

The meeting with Socrates was fundamental. Plato was particularly affected by the fact that his teacher was accused of impiety and corruption of young people by the Athenian democratic government. Socrates was sentenced to death in 399 BC.

After Socrates' disappearance, his disciples, including Plato, left Athens to take refuge in Megara.

Starting from 395 BC Plato began writing the first "dialogues". In these writings he addresses the cultural problem represented by the figure of Socrates and his unjust condemnation. Plato bitterly criticizes the function of the sophists. The Sophists were "masters of virtue" who were paid to bestow their teachings, so they were called "cultural prostitutes".

In 387 BC Plato returned to Athens; and bought a park dedicated to Academo. In this park he founded a school which he called the Academy.

The Academy had its roots in science and in the particular teaching method called dialectics. The teaching took place through debates, in which the students themselves participated, directed by Plato or by the older students. Sometimes conferences of distinguished people passing through Athens were held in the Academy. Plato died in 347 BC

Platonic philosophy

According to Plato, the human being is made up of two parts. The first part consists of a body hopelessly linked to the sensitive world. Therefore, the first part follows the same fate as everything belonging to this world, such as plants and animals.

The second part of the human being is the soul, composed of consciousness and reason. This part is not made of matter and therefore is immortal, because it cannot fall into dissolution.

Plato affirms the existence of the World of Ideas. This world exists as the world of material things. The world of ideas is the privileged place for the existence of the soul and can be observed by man through reason.

Souls await in the World of Ideas waiting to take up residence in a body, and perfect knowledge of Ideas matures in this expectation. When, however, souls descend into a body, they forget this knowledge, or rather, they retain only a faint memory of it. Therefore, every form of learning that develops in human life is not a fresh acquisition of information, but a process of reminiscence of things already known, even if forgotten.

The conceptual basis of Plato's philosophy is developed in various dialogues such as Menone, Phaedon and Theetetus. In these works he denies that the search for knowledge is impossible, as claimed by the Eristi in their assumptions, summarized as follows:

- If you don't know what you're looking for, if you find it, you wouldn't recognize it as the goal of your search.
- If you already know what you are looking for, your search does not make sense.

Plato solves the problem by opposing his philosophy of Ideas to these assumptions. The object of the research is only partially unknown, because the soul of each man contemplates the original Ideas before birth, but then forgets them. Consequently, the goal of man's search is knowledge that is already present in him, even if forgotten.

The myth of the cave

For Plato, Ideas are perfect and immutable, and pre-exist with respect to the body, therefore they cannot derive from the senses (sight, hearing, smell, and so on) which are imperfect instruments of the body. To explain this concept he uses the myth of the cave. This is a place of imprisonment for men forced to look at a deceptive reality, the shadows reflected on the back wall. The real images are found outside the cave, in the light of the sun, that is, in the light of reason and knowledge not mediated by the senses; unfortunately, the prisoners don't know.

Plato proposes this which is the most famous of his myths in Book VII of his work Repubblica and exposes it through a dialogue between Socrates and the disciple Glaucone.

> [Socrates] And now, I said, let me show in a figure how far our nature is enlightened or unenlightened: -- Behold!
>
> human beings living in a underground cave, which has a mouth open towards the light and reaching all along the cave;
>
> here they have been from their childhood, and have their legs and necks chained so that they cannot move, and can
>
> only see before them, being prevented by the chains from turning round their heads. Above and behind them a fire is
>
> blazing at a distance, and between the fire and the prisoners there is a raised way; and you will see, if you look, a low
>
> wall built along the way, like the screen which marionette players have in front of them, over which they show the
>
> puppets.
>
> [Glaucon] I see.
>
> [Socrates] And do you see, I said, men passing along the wall carrying all sorts of vessels, and statues and figures of
>
> animals made of wood and stone and various materials, which appear over the wall? Some of them are talking, others
>
> silent.
>
> [Glaucon] You have shown me a strange image, and they are strange prisoners.
>
> [Socrates] Like ourselves, I replied; and they see only their own shadows, or the shadows of one another, which the
>
> fire throws on the opposite wall of the cave?

[Glaucon] True, he said; how could they see anything but the shadows if they were never allowed to move their

heads?

[Socrates] And of the objects which are being carried in like manner they would only see the shadows?

[Glaucon] Yes, he said.

[Socrates] And if they were able to converse with one another, would they not suppose that they were naming what

was actually before them?

[Glaucon] Very true.

[Socrates] And suppose further that the prison had an echo which came from the other side, would they not be sure to

fancy when one of the passers-by spoke that the voice which they heard came from the passing shadow?

[Glaucon] No question, he replied.

[Socrates] To them, I said, the truth would be literally nothing but the shadows of the images.

[Glaucon] That is certain.

[Socrates] And now look again, and see what will naturally follow if the prisoners are released and disabused of their

error. At first, when any of them is liberated and compelled suddenly to stand up and turn his neck round and walk and

look towards the light, he will suffer sharp pains; the glare will distress him, and he will be unable to see the realities

of which in his former state he had seen the shadows; and then conceive some one saying to him, that what he saw

before was an illusion, but that now, when he is approaching nearer to being and his eye is turned towards more real

> existence, he has a clearer vision, -what will be his reply? And you may further imagine that his instructor is pointing
>
> to the objects as they pass and requiring him to name them, -will he not be perplexed? Will he not fancy that the
>
> shadows which he formerly saw are truer than the objects which are now shown to him?

The myth of the cave represents a perfect analogy between the condition of prisoners and our condition. Analyzing the symbolism of the myth we can establish some parallels:

- Chained prisoners represent ourselves.
- The interior of the cave is our empirical world, that is, the daily world of our experience.
- The chains that imprison us are the senses. The senses are absolutely imperfect tools that force us to see a false reality.
- The shadows that appear at the bottom of the cave are dreams, fears, fantasies and illusions. These are built on the overall falsehood of the situation.
- Freeing oneself from the chains in order to see reality and sunlight means freeing oneself from the conditioning of the senses.

This conditioning is not philosophical but physical. We can hear and see only within the narrow limits of possible frequencies, we can taste within the limits of sweet and salty. Therefore, we can only have limited information on objects. We can evaluate their hardness and heat in an absolutely gross way. At the same time, we cannot understand anything about the finer essence of the molecules and atoms that make up matter.

The release of the senses from the chains cannot be physical, because we cannot force the senses to function differently. Moreover, we must be grateful to the senses for their functioning, because they enable us to interact with the macroscopic reality in which we live.

Therefore, liberation must be intellectual. We must build in ourselves the awareness that there is a reality "out there". The senses do not allow us to see this reality. However, this reality can be experienced with reason.

At the level of consciousness, we can take note that all the truths denied by the senses, the spirituality, the soul, the psychic dimension of ideas and an ordering Intelligence of the cosmos are right out there and can break into our consciences illuminating them with the light of truth . This can happen despite the fact that, physically, in everyday life, we remain chained by the prison of the senses.

Prisoners have never seen anything but their shadows, so they think shadows are reality. However, a prisoner breaks free and looks around. He sees the objects and tools that pass over the wall and understands that they are more real than the shadows. He comes out of the cave and remains dazzled by the sun. He senses that everything is more true outside the cave. He wonders where the light comes from, and realizes that the light gives meaning to the whole. Light represents knowledge and the idea of good and beauty.

However, Plato does not assign a totally negative value to shadows. The path leading out of the cave begins right from the shadows. Shadows are related to real objects. According to Plarone, above all there are the Ideas, from which the objects of the sensitive world derive. The Ideas are the Archetypes, that is, the models. Real objects, which are reflected as shadows for prisoners, are shaped on these models.

All of Plato's philosophy presupposes the existence of a purely psychic or spiritual higher level, the World of Ideas, and has many similarities with the theory of archetypes and the collective unconscious developed many centuries later by Carl Jung.

All of Plato's philosophy presupposes the existence of a purely psychic or spiritual higher level, the World of Ideas, and has many similarities with the theory of archetypes and the collective unconscious developed many centuries later by Carl Jung.

Figure 4 - Some personalities who have stated theories contrary to materialism (photo from the web).

From Plato to vitalism

Vitalism developed between the eighteenth and nineteenth centuries, in the wake of neo-Platonic ideas and Renaissance philosophy. According to this current of thought, Plato's Ideas are extended to all of nature. Ideas are a constitutive part of every single organism and everything that exists. Therefore, the whole cosmos is animated by an intelligent principle, the Soul of the world. (in Latin *Anima Mundi*).

This philosophical term indicates the vitality of nature as a whole. Nature is considered similar to a single living organism. Hence, the Anima Mundi is the unifying principle of the cosmos. From this principle, individual organisms take shape. All the organisms are articulated and differentiated each according to their own individual specificities, but are linked together by the common Universal Soul.

Gottfried von Leibniz attributed life and thinking ability even to inert matter. This, of course, was in stark contrast to materialistic determinism, which never accepted the possibility of a profound unity of all nature.

A great supporter of vitalism in the 20th century was Henri-Louis Bergson (1859-1941), French philosopher and Nobel prize for literature in 1927. In his writings, in controversy with materialistic determinism, he reaffirms that biological life, such as consciousness, it is not a random aggregation of chemical compounds that reproduce, but it is a perennial and incessant creation.

Bergson argues that at the base of everything there is a vital impulse, which pushes the material towards increasingly complex realizations. This momentum expands in different directions and with different intensities. In this way he explains the existence of the world of plants separate from the world of animals. In Bergson's conception the materialist theses of mechanism and determinism no longer have meaning. The evolution conceived by Bergson starts from the assumption that nothing is fixed or determined, but everything is reality in motion. Evolution is both the subject and the object of itself, because it follows a process that has a single origin. This process takes place in a continuous impulse towards life, therefore the Bergsonian process represents a creative evolution. In this context, matter is only the point of arrival of the development of a certain way of creative impetus. So, according to Bergson, matter is a manifestation of the spirit.

Bergson also condemns attempts to uncover the mystery of life by analyzing the individual parts of a living being. In this sense, he shares the statement of Mephistopheles in Faust:

> "To understand and describe a living reality, one proceeds first by killing and dismembering it.
> Then you only have inert fragments in your hand which - unfortunately - lack any reference to life ".

The mystery of life can never be understood by dismantling a living being into pieces. In fact, materialists would like to do it, in imitation of a child who takes apart the toy to understand how it works. With the toy you can do it, because the mystery of its functioning lies in some spring or gear, therefore the mystery is located inside the toy. The mystery of life, on the other hand, is not inside the creature: it comes from the outside, it is part of the Soul of the world.

From Hans Driesch to Carl Jung

A figure of particular interest in the field of vitalism is that of Hans Driesch, (1867-1941) German philosopher and biologist, a pioneer of research in the field of embryology. Between 1891 and 1900 he conducted research on the sea urchin at the Zoological Station of Naples. At first he tried to explain the development of embryos based on mechanistic theories, but he soon had to acknowledge the insufficiency of those theories. He espoused the thesis of a vital non-material principle inherent in organisms but coming from outside them. This principle coincides with the purpose of nature. The principle hypothesized by Driesch is the cause of the development and evolution of living beings, both as a species and as individuals in the species, each with its own particularities.

Drawing from Aristotle's studies, Driesch called this principle "*entelechy*". This term indicates a reality that contains in itself the final goal towards which it tends to evolve. Applied to a species or even to an individual, entelechy presupposes the existence of a "place" where this information resides.

It is a "place" external to the physicality of the individual. However, individuals can draw information from this "place".

Driesch's theories were particularly fascinating to Carl Jung. In his work published in 1952, " *Synchronicity as a principle of acasual connections* ", with which he officially announced his theory of synchronicity to the world, the Swiss psychologist acknowledges that he has been influenced since 1925 by the reading of Driesch's writings. Thus writes Jung in a note of his work:

> "The psychic purpose is based on a" pre-existing "sense that becomes problematic only later, when it becomes an unconscious adaptation. In this case, in fact, one must suppose a kind of "knowledge" prior to all consciousness. Hans Driesch also came to this conclusion. "

Jung's reflections on synchronicity had already started in 1916, but the first explicit formulations of the theory took place around 1925, coinciding with the findings with Driesch's theories.

Ernst Schumacher's "degrees of being".

In his volume "Epochal winter. Current considerations" Hanspeter Padrutt takes up and comments on a special ranking formulated in the last century by Ernst Friedrich Schumacher, on the different degrees of being (fig. 5).

Ernst Schumacher was an influential German philosopher and economist, born in Bonn in 1911 (fig. 4). He emigrated to England before the Second World War in order not to live under the Nazis: despite this, when the war broke out, he was considered an "alien enemy" and was interned in an isolated farm in the United Kingdom. During this time he met the famous economist John Keynes. Keynes appreciated some of his works and obtained his release. In this way Schumacher was able to contribute to the economic and financial recovery of the British government. As a result, Schumacher earned a good position at the University of Oxford.

Ernst Schumacher became interested in the problems of coal mining, and became known for his critique of Western economies and his proposals for adopting more humane industrial technologies.

In 1977 he published "A guide for the perplexed". In this essay he criticizes scientific materialism and delves into the exploration of the nature and organization of the knowledge of being. In this little "guide to living in the world", Schumacher renounces the paradigms of classical science and invites the reader to refer to himself and take possession of his own life and the meaning of existence, to "live" subjectively and not "be lived "objectively.

In addition to economics and philosophy, his great passion was gardening to which he dedicated himself with passion. He died in 1977 while traveling in Switzerland.

He transferred part of his philosophical research into a table. In this we note his training as an economist. Indeed, Schumacher expresses his idea of the different levels of consciousness and knowledge in an unusual way that is more reminiscent of mathematical formulas rather than philosophical concepts. Undoubtedly, this method gives great clarity to the concepts.

Psychologist and philosopher William James (president of the Society for Psychical Research from 1894 to 1895) is known for his claim:

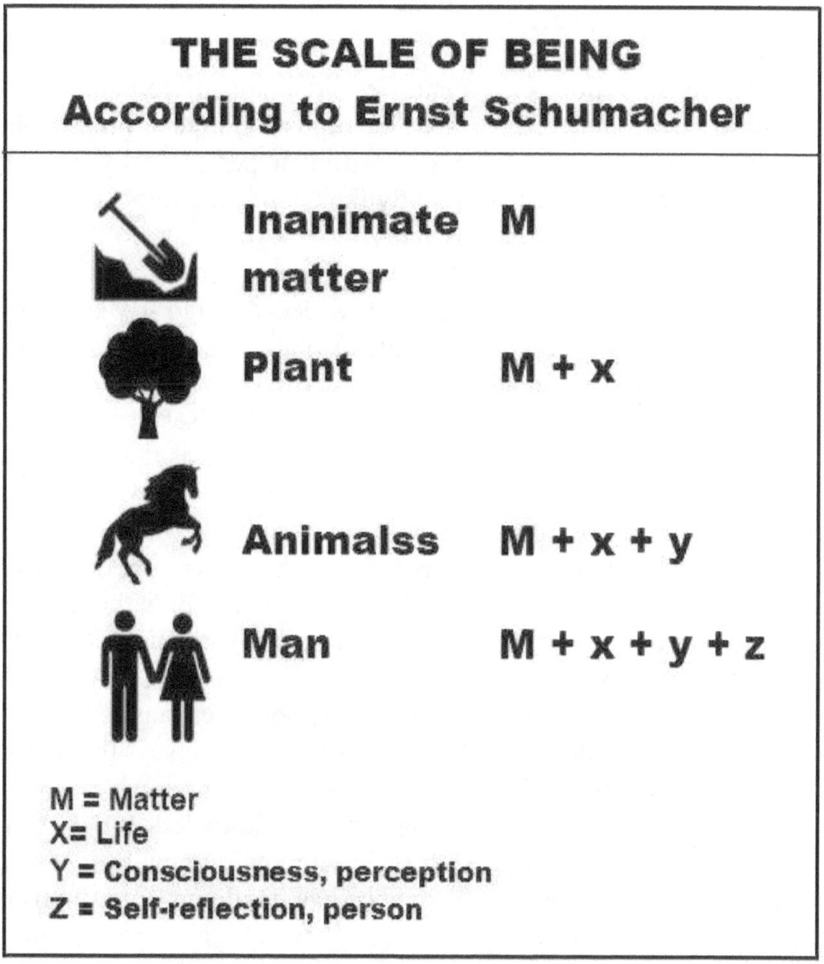

Figure 5 - According to Ernst Schumacher, an influential German philosopher and economist, each level of reality implies the existence of a higher level.

"What we know is only a small drop in the great ocean of ignorance".

This observation is two centuries old, but it still remains true and current, especially today. In 1835 Auguste Comte, a French philosopher considered the father of Positivism, peremptorily stated that we could never in any way study the chemical composition of the stars. Today this is done normally.

In the scientific field, today more than yesterday, it is quite clear that we know very little. Only a few decades ago it was believed that atoms were the smallest part of matter and that this was not further divisible; instead, today, new particles continue to be discovered in accelerators and quantum physics raises serious doubts about what particles are, coming to the hypothesis that they do not exist at all but consist only of vibrating energy.

The horizon of human knowledge goes further and further: the theory of relativity has just upset our certainties about the fixity of time and space, and we are already ascertaining how that universe, which we thought we knew so well, is unknown to us. at least 90%. Dark matter could be worth just 90% of the universe. As for dark matter, it resembles the phoenix: of the poet Metastasio: "che ci sia ciascun lo dice, dove sia nessun lo sa". (everyone says that there is, no one knows where it is)

Every day we conquer a small space of knowledge, only to realize then that that space is an open window on new immense and unknown territories. Today science has hundreds if not thousands of windows open onto panoramas of which only the existence is certain, while the contents remain completely unknown.

We know we are made of matter. We are an aggregation of molecules, which in turn are aggregates of atoms, which in turn are aggregates of elementary particles. However, what place does the spark of life have, where does it reside? Where does the organization of these aggregations of matter reside? Our body is a vehicle, an extremely efficient machine; but the driver, where is he?Did it all come from chance? The matter, after having miraculously aggregated in our body, has it also acquired the ability to guide itself?

In 1981 two renowned astronomers, Fred Hoyle and Chandra Wickramasinghe, wanted to calculate how likely it was that life was born just like that, by chance. They concluded that there would be a probability equal to that of a hurricane that, by unloading its power on a landfill of metals and moving them in the whirlwind of the winds, would be able to form a Jumbo Jet ready to fly. Perfect. But even if this were to happen miraculously, where would the pilot capable of taking off in flight come from?

Despite the best efforts of the materialists, it is evident that matter alone cannot go beyond a certain limit. Behind the perfection of the universe there must be an intelligence, a higher consciousness that guides everything.

Humanity is waiting for science, as a near and necessary goal, to finally set itself the goal of renouncing materialism and the search for a non-material level, a "Universal Consciousness" independent of time and space, a "Field of energy and information "containing the" Project ", the" Instruction Booklet ", all the knowledge from which individual consciences already draw, unwittingly, today.

Our individual consciousness is part of a larger consciousness, from which it draws information and wisdom.

My liver probably doesn't know it's part of a larger whole, which is my body. Likewise, I am part of a larger whole, without knowing it. However, the larger whole of which I am a part knows me and takes care of me.

In this sense, Schumacher proposes in his table different stages in which a part of the universe is part of a larger whole, even without knowing it; the larger whole, in turn, includes the parts that are lower in the scale and adds something more (fig. 5).

The lowest level (M) is that of inert matter which can be weighed and measured. It is part of the higher level, that of the plant world. This level includes M but adds x, which stands for life.

Therefore, if the mineral level is only M (matter) the next level, the vegetable one, is M + x, matter + life. But this level is, in turn, part of a higher level, the third.

The third level, relating to the animal kingdom, includes the lower (M + x) but adds y, that is, consciousness and the ability to perceive. Thus, the third level becomes M + x + y.

The whole third level is then part of the fourth, that relating to man, which adds z, that is, the self-awareness of one's "Self". Therefore, in the degrees of being man is M + x + y + z.

Schumacher stops his ladder here. But if each step is naturally part of a higher step, what will the human level become part of?

Why shouldn't there also be a higher level, which includes M, x, y, z but also "something more"?

The most natural answer is yes. There is the fifth level, that of "Universal Consciousness", which adds value.

Not knowing how to call and define this value, we call it "infinite". Its symbol is ∞.

The fifth level of the scale would become M + x + y + z + ∞.

The scale of the degrees of being, thus completed, makes man part of a universal Consciousness that includes the whole universe: Matter, life, consciousness, self-awareness, energy and infinite information.

In this way the immaterial part of ourselves, our personal pilot, would be united with the great universal pilot from which he would receive the patent. Our personal conscience would be in continuous dialogue with the universal Consciousness and finally we would know where our deep intuitions come from, our a priori knowledge, our noblest feelings and absolutely extraneous to that animality in which materialism would like to force us.

Moreover, throughout the history of humanity, man has always intuitively believed in something higher that he has called by the names of infinite divinities. There is certainly a Cosmic Intelligence that is beyond our understanding. We are continually affected by it.

It seems that man has an innate need to believe in something greater. After all, even the conviction that there is nothing greater is still a belief, a faith.

Dreams that suggest inventions and successes.

The affirmation that messages useful for the development of individuals or even humanity derive from Universal Consciousness is confirmed by many famous personalities of the arts and science who claim to have received the inspiration for their creations in a dream. History is full of problems solved at dawn. For example, Friedrich Kekulé, a German chemist whose name is associated with the definition of the structure of benzene and aromatic compounds, dreamed of a snake biting its tail and upon awakening he realized that the benzene molecule was hexagonal.

Mendeleev's Periodic Table.

In Russian scientific circles it is claimed that Dmitry Ivanovich Mendeleev came to determine the composition of the Table of Elements thanks to a dream. The famous chemist (1834–1907) wanted to rationally organize the 65 chemical elements known at his time. He had guessed that the organizational model had to be linked to the atomic weight, but he could not determine the exact relationship. According to the Russian chemist B.M. Kedrov, in his essay "On the question of scientific creativity", Mendeleev stated:

> "In a dream I saw a flat surface, where all the elements were arranged in the required order. After waking up, I immediately drew that arrangement on a sheet of paper".

Thus was born the periodic table that we still know today. Mendeleev noted that the arrangement seen in a dream was so accurate that it revealed that some elements had been measured inaccurately.

The Google search engine.

Larry Page, co-founder of Google, says that the idea of a search engine with the characteristics of Google came to him during a "vivid" dream. After talking about it with his friend Sergey Brin, they gave life together with Backrub, the precursor of Big G.

Paul McCartney and Yesterday.

Paul McCartney tells his biographer Barry Miles that one morning in 1965 he woke up with music in his head. He got up and played the melody that would become "Yesterday" on the piano. Today this dream born song is in Guinness Book as the most recorded song.

The transmission of signals in the brain.

In the early 1900s, scientists mistakenly believed that the transmission of signals in the brain took place electrically. Instead, a dream of Otto Loewi suggested the exact theory. Loewi was a lecturer at the University of Graz in Austria, but moved to the United States during the Nazi regime. Here he taught at the University of New York, the city where he died in 1961. His studies focused on the pharmacology of substances that act on the autonomic nervous system, both on the sympathetic and on the parasympathetic, discovering that the mediators active in the functioning of the nerves and muscles are chemical and not electrical.

Loewi came to this discovery after a dream. He woke up and scribbled some notes on a piece of paper. Upon awakening he realized that he had written the foundations of the theory that would change the history of medicine: that information is transmitted chemically, through active chemical mediators. Thanks to this discovery, Otto Loewi won the Nobel Prize for physiology in 1936.

Niels Bohr and the atom.

In an article written at the Massachusetts Institute of Technology (MIT) by Edwina Portocarrero with some of her collaborators we read the following:

"Niels Bohr (1885–1962) developed the atomic model based on a dream of him. In the dream, Bohr was sitting on the Sun while all the planets, turning around, made hisses and moved as if on strings ».

Indeed, the current atomic model consists of a central nucleus and a series of electrons revolving around it. Portocarrero's article was dedicated to dreams and was entitled "Lessons on the pillow: discontinuous interface for the recall and reproduction of dreams".

Mary Shelley and Frankenstein

In May 1816, a group of writers were staying in a hotel in Geneva. Among these writers were Lord Byron, his lover Claire Clairmont with her sister Mary Shelley, and other people who love letters. The day was rainy, and they decided to overcome boredom by reading German ghost stories. Byron then proposed to compose a ghost story themselves: everyone agreed. Mary Shelley had no idea how to start, but in the course of the night she saw in a dream the moment when the mad scientist created Frankenstein. The next morning she set to work writing the chilling sequel, and the book was as successful as we know it. What would horror literature be today without Frankenstein?

"Misery doesn't have to die"

This is one of Stephen King's most popular books, from which a movie was also made. The author himself says that the idea of the plot, based on a fan who holds her favorite writer prisoner to rewrite the ending of a story, came to him during a dream on a plane.

Elias Howe's sewing machine.

The invention of the sewing machine is attributed to Elias Howe (1819–1867), who had the merit of greatly improving some existing projects and of patenting its improved version. For a long time Howe was stuck with his project: in fact, Howe did not know where to place the eye of the needle.

Cause this problem the work was stopped and he was falling into poverty. His original idea was to follow tradition and place the eye on the upper end. However, in doing so, the thread always got tangled. He had never thought of placing the eye near the tip of the needle.

In 1845 Howe had a strange dream. Some cannibals had captured him and stabbed him with their spears. His expert eye noticed that the spears had a hole next to the tip. Said and done, he applied this idea to his invention and the sewing machine was born.

Wolfgang Pauli's dreams.

The Nobel laureate physicist Wolfgang Pauli repeatedly stated that he received the inspirations in a dream, useful for the formulation of his scientific theories.

A dream that occupied Pauli throughout his life was what he called the "World Clock". In the dream Pauli saw two concentric wheels turning non-concentric. Unfortunately, the wheels were connected by an impossible central axis. At the time of the dream Pauli was in analysis at Jung's studio, and he understood that the image contained a message to be transposed into his scientific research: for Pauli the two wheels represented the psyche and matter, which needed to be unified into one. scientific physical theory. Unfortunately, although he pursued this goal as a lifelong mission, Pauli was unable to find a way to reconcile scientific materialism with spirituality. Indeed, modern science rejects any reference to the psyche or the spirit as leprosy.

From Plato to Berkeley.

In the myth of the cave Plato suggests that all things, as we see them, are nothing more than a gigantic illusion, but he fails to explicitly state that things do not exist. This concept, however, is enunciated several times in human thought, and becomes the backbone of George Berkeley's philosophy (fig. 6). Theologian and Irish Anglican bishop, Berkeley, with John Locke and David Hume, was one of the three great British empiricists. According to Berkeley, the only objects of human knowledge are ideas.

What we define as a thing is nothing more than a set of ideas. For example, an apple is nothing more than the set of a certain color, a certain flavor, a certain smell, a certain shape, a certain consistency.

All the objects that we believe exist (because the senses make us believe they exist) in fact do not exist at all: they have no consistency, they have no body, they are only ideas that, conveyed by the senses in the brain, we configure by giving them a false impression of materiality.

We do not know objects at all: we only know the ideas of objects that the senses transmit to our brain. We live as in a dream, like the prisoners of the Platonic cave who observe the shadows believing them to be real. We have spatio-temporal perceptions relating to presumed objects which, however, do not exist..

The Berkeley tree.

It is useful to repeat an ancient question: if a tree falls in a wood and there is no one around, does it make noise? This question was proposed by George Berkeley.

Berkeley was born in Kilkenny, Ireland on March 12, 1685 and died in Oxford on January 14, 1753. He graduated in 1707, completed his undergraduate studies in mathematics, logic, classical literature, theology and philosophy at Trinity College Dublin. He immediately deepened his knowledge of the great thinkers of his time by reading the works of Locke, Gassendi, Descartes, Newton, Descartes, Hobbes and others, thus giving form and direction to the evolution of his thought, such as to be able to enunciate the cardinal principle of his philosophy: immaterialism.

Figure 6 - George Berkeley was an Irish Anglican philosopher, theologian and bishop. He states that the only objects of human knowledge are ideas; things in themselves do not exist, they are just a set of ideas.

In 1709 he published the Essay of a New Theory of Vision and the following year the Treatise on the Principles of Human Knowledge. In 1713 he also wrote the Dialogues between Hylas and Philonous.

In 1728 he married Anne Forster, eldest daughter of John Forster, President of the Irish House of Commons. Having benefited from a conspicuous inheritance, his spirit of fervent religious led him to want to leave, with his wife, to the new lands of America where he intended to carry out missionary activity by founding a College in the Bermuda Islands.

While waiting to receive the promised funds, Berkeley moved to America and settled in Rhode Island, where Henry, his first child, was born. Unfortunately, the philosopher was then forced to abandon the initiative because the Queen did not keep the funding promise made in the first place. Back in Ireland, he was appointed bishop of Cloyne (for the Irish Church).

Berkley's philosophical thought is all centered on idealism that he strongly contrasts with the "Enlightenment" that peeps out between the end of the 17th century and the beginning of the 18th century. Berkeley rejects materialism that degenerates into skepticism and atheism. On the merits he composed "Alcifrone", a polemical dialogue against the free thinkers of the time.

Berkeley argued that objects exist only as perceived. Therefore, if a tree falls in a forest where no one witnesses the fall, it produces no noise.

From the point of view of our common sense we could object that obviously the falling tree certainly produces a noise, because the breaking of the woody fibers and the collision with the branches of other trees cause the air displacements typical of sound waves.

Apparently, however, the same science has intervened expressing a favorable opinion to Berkeley. In fact, one of the oldest and most prestigious scientific popularization journals, Scientific American, founded by Rufus Porter in 1845, took the field in 1884 and rationally demonstrated that, indeed, a falling tree makes no noise, if there is no one who can hear the noise.

In practice, according to the magazine, it is true that the fall produces vibrations in the air, but these can only be detected with an auditory system that, through the internal parts of the ear, can translate them into sounds. Therefore, in the absence of an auditory system capable of detecting noise, this is not there. Flawless explanation.

In this way we solve a question that is sometimes mistakenly attributed to Zen philosophy: if the tree falls when there is no one to hear it, it makes no noise.

Returning to Berkeley, to reinforce his thesis he coined the saying "Esse est percipi" (To be means to be perceived). The whole being of an object consists in its being perceived, and nothing else. It is the summary of his immaterialist theory, according to which things are nothing more than a series of ideas, and begin to exist only when they are perceived.

This is not a particularly surprising revelation for a quantum physicist. As we will see, at the level of elementary particles, these change their behavior if there is someone who observes them, or not: moreover, depending on how they are observed, they can change the shape in which they appear.

After centuries of ironic giggles, Berkeley's theories begin to take on a consistency firmly attached to reality.

Of the same kind as the falling tree is a more technological dilemma: if we turn on a light bulb in an empty room, where no one can observe it, does it emit light? Normally you would answer yes; except that, light is nothing more than a perception of the nervous system, so if there is no system in the room capable of perceiving light, that is, an eye, there is no light. Light begins to exist when someone can see it: it does not have an existence of its own, it is only the elaboration of a visual system. In fact, by closing the switch, a form of energy flows and changes into another form of energy, which however we do not know. It takes on the appearance of light only when someone observes it.

Similarly we would say: what does that tempting dish that was served to you taste like? Nobody. Food will begin to taste when you submit it to your palate. In itself, flavor is a quality that is completely unrelated to the food you are eating.

At this point it seems reasonable to ask ourselves if it is we who adapt to reality, or if it is not reality that adapts to us.

The universe exists because we look at it.

Let's go back to the theme of light. If it is true that the light bulb in a closed room does not emit light because no one can see it, in the same way we can say that, if we were not there to look at the universe, the universe would be dark. The stars would not shine, the sun itself would not emit light, if there was no one to look at them.

For the universe to be what it appears, someone must be there to look at it. Without eyes capable of transforming the photons emitted by the stars into light, the universe would be dark.

In the second verse of the Bible (Genesis 1, 2) the author begins the account of creation with these words:

> "And God said, "Let there be light," and there was light. And God saw that the light was good, and He separated the light from the darkness.... "

God saw that the light was good. After creating it, he saw it and he admitted it was a good thing. Light becomes good when God sees it. Not only that, when God creates man, he creates him "in the image and likeness of God". In this way, man too can see the light so that creation remains a "good thing" insofar as it is a "thing looked at".

The main objection to this should come from the theory of evolution, according to which all living beings derive from a single ancestral organism from which all current biological diversity would have developed, and for each living being the most suitable organs would have developed. ensure its survival. All organs (hands, legs, tail, feathers, ears, skin, and so on) would be the result of the progressive improvement of pre-existing organs, or in any case developed by the ancestral ancestor.

Therefore, a fin can turn into a leg to accompany the passage of fish to land animals: that is, an organ developed in a given condition as useful for survival in that condition, can evolve into a different organ to be useful in a different condition. . This theory is credible for almost all organs, but not for the eye.

The evolutionary contradictions of the eye.

If there is a problem that torments Darwinists, it is precisely that of the evolution of the eye. In fact, Darwin himself, in his work "The Origin of Species" states:

> "To suppose that the eye, with all its inimitable mechanisms [...], could have been formed by means of natural selection, it seems, I confess freely, absurd in the highest degree."

The first apparently effective attempt to explain the gradualistic development of the eye dates back to 1994: it is the famous work by Nilsson and Pelger "A pessimistic estimate of the time required for an eye to evolve". According to these scholars, the eye evolved from an initial layer of light-sensitive cells, thanks to which the primordial organism was able to direct itself towards the light source. This confirms, in any case, that the need to see was already inherent in the first organism.

Subsequently, the evolutionary processes would have caused the bagging of these photosensitive cells in a kind of hemispherical well; when the cockpit became deep enough, a narrowing diaphragm appeared on the edges. Eventually, a denser area capable of acting like a lens would have formed in the recess already filled with transparent substance.

Unfortunately, the evolutionary chain described in these passages is by no means a gradual and continuous evolution, and does not correspond to what should be minimal successive modifications of the same characteristic, as it includes the addition of other completely independent organic structures: a " proto-pupil "and a" proto-crystalline ".

In fact, the eye does not represent the evolution of a single primordial organism but requires the addition of different and sophisticated parts, such as lens, regulatory muscles, sensory cells and innervations. A lens without a retina, as well as a retina without a lens, is not functional. Yet, retina and lens are not part of the same evolutionary process because they have different tissue qualities that imply different origins both in embryogenetic development and in evolution.

This undoubtedly confirms that the eye is the product of an evolutionary convergence of different parts in a single integrated function: therefore, if we want to speak of randomness, it would be a chance elevated to the nth degree.

In conclusion, it is inconceivable that a device as specialized as the eye could arise from random developments, all initially without specific evolutionary advantages, which then miraculously integrated into the visual function.

We can imagine the development of a primordial eye, devoid of retina and crystalline, only if we understand it as an obligatory passage towards a future function: that is, as part of a project that already predicted what its use would be in subsequent eras. Certainly not a blind evolution, but a project guided and assisted by a universal intelligence.

When things are not what they seem.

A human being is part of a whole that we call "universe", a part limited in time and space. He experiences himself, thoughts and feelings as something separate from the rest, in what is a kind of optical illusion of consciousness.

This illusion is a kind of prison that limits us to our personal desires and affection for the few people closest to us. Our task is to free ourselves from this prison, by enlarging our compassion in concentric centers to embrace all living creatures and all nature in its beauty.

(Albert Einstein, physicist, Nobel laureate)

The strange case of the two presidents.

In recent times it is very fashionable, in the gossip press, to look for the strange coincidences that unite celebrities. In particular, the search for facts, events and circumstances that see Kate Middleton and Charlene of Monaco as protagonists is going crazy.

The coincidences are numerous, but they don't seem particularly significant. For example, there is a lot of evidence that both Kate and Charlene have no blue blood. They were both born in January, Kate on the 9th and Charlene on the 25th. Still as far as dates are concerned, Kate and Charlene both got married in 2011.

From a more personal point of view, the two noblewomen are both sporty and love to attend any type of competition, engaging in various disciplines. Plus, they both have to deal with two rather "bulky" mothers-in-law, Lady Diana for Kate and Grace Kelly for Charlene.

Certainly, the narrative of the coincidences that bind two of the most famous presidents of the United States together is more intriguing. We are talking about John Fitgerald Kennedy (fig. 7) and Abraham Lincoln (fig. 8). The most relevant coincidences are related to the modality of their death. In fact, both presidents were killed in very similar circumstances, Lincoln on April 14, 1865 and Kennedy on November 22, 1963. Here are some of the more well-known coincidences with respect to their killing:

- Both were killed on Friday, with a gunshot to the head.
- Both had their wife beside them, who was not hit.

Kennedy was killed while in a Ford Lincoln Continental limo; Lincoln while attending the show "Our American Cousins" in the Ford Theater in Washington.

- At the time of the attack, both Lincoln and Kennedy were accompanied by a couple: in both cases the man, who was next to the president, was injured.

Figure 7 - An incredible series of coincidences links two of the most famous presidents of the United States, Abraham Lincoln and John Fitgerald Kennedy, both victims of attacks.

- Both killers (John Wilkes Booth for Lincoln, and Lee Harvey Oswald for Kennedy) had three names for a total of fifteen letters.
- Both killers died under unclear circumstances before being tried; however, their deaths occurred at roughly the same clock time as the victims, and both died from gunshot wounds..

There are also coincidences regarding private life or political career.

- Both had seven-letter surnames, and had been baptized with the same name as their respective grandparents. Furthermore, both were second-born of their own families.
- Both of them had fought in the war; Lincoln in the Black Hawk War, and Kennedy in World War II.
- Both of them lost a child during their presidential term. Lincoln lost William Wallace, and Kennedy lost Patrick Buovier.
- At the time of death they both had four children.
- Both were first elected to Congress in the 46th of their century; likewise in 1956 they were candidates for the vice-presidency, but were not elected because their party lost the elections.
- They were both engaged in defending the rights of the black population, Lincoln in relation to the abolition of slavery and Kennedy in the promotion of civil rights.
- Both were succeeded by a vice president named Johnson: Andrew Johnson for Lincoln and Lyndon Johnson for Kennedy.

If that's still not enough, those who want can find hundreds of other coincidences that connect the life of the two presidents, by consulting the website (in English): https://www.near-death.com/: https://www.near-death.com/

The explanations of the skeptics.

While half of the world strives to study cases like these to add other coincidences to the list, the other half does their utmost to demonstrate that some of the coincidences presented would be false and would try to use some inaccuracy to deny it all.

Figure 8-Abraham Lincoln, victim of an attack on April 14, 1865. Many circumstances of his life and his death are incredibly coincident with similar ones involving President John Kennedy

Then there are some general objections that are good for every occasion, such as that it is not so difficult to find coincidences between two facts chosen at random, if you go to examine all the details.

According to the law of large numbers, if you evaluate the life and work of two famous people in detail, it is not difficult to build a list of coincidences. If, for example, we referred to Masaniello and General Lamarmora, it is likely that both of them liked Neapolitan pizza and both of them wore 44 on the foot, but I don't know what other really significant coincidences could be found.

The deception of the senses.

But why does science offer so much resistance to the paranormal? In fact, Western science has spent several centuries of its life not believing it, and today, as cracks begin to appear in this granite certainty, it lets itself be carried away by habit. In fact, it is difficult to give up conveniences, consolidated positions, careers built on the affirmation of materialism.

Moreover, we ourselves, who are the water on which the ship of science glides, are so addicted to these positions that we take them for granted, despite the fact that every day we have to deal with the paranormal or with extrasensory perceptions. We are now used to believing only in what can be touched, weighed and measured, so not even the evidence is able to eradicate the dogmas of materialism from our mentality.

In the anxiety of giving importance only to the tangible (perhaps because only matter can be monetized) we forget that, in most cases, the most important things cannot be held in hand to turn them into money.

The main source of our well-being are thoughts, positive moods, feelings and all activities related to the spirit or psyche that are able to give profound meaning and motivation to our existence or, on the contrary, manage to make it appear useless. and no future.

The facts that happen around us are shaped by our way of interpreting them, by our optimism or our pessimism, by our gradient of trust and hope or by that of despair. There is no objective reality around us, things are not like that in any case, but they change according to our interpretation, that is, through the filter of our conscience and our reason.

In this sense, what we perceive with the five senses, that is sensory reality, is always the result of an extrasensory processing.

To say that extrasensory abilities do not exist is like considering ourselves weathervanes exposed to every wind that passes, without any possibility of resistance.

Instead, thanks to the extrasensory abilities that cannot be measured and weighed but certainly there are, such as courage, an optimistic vision of the future, the awareness that life is made up of ups and downs and every low is always naturally followed by a high, the trust in ourselves and perhaps also in our neighbor, mercy and kindness, therefore the ability to help and be helped, all this makes the shafts of our weathervanes as solid as rocks. Any wind can blow as hard as it wants, without ever being able to eradicate these beliefs.

What we perceive with our five senses, in times of storm and in those of calm, is always a mediated and medable truth.

This is valid not only on the psychological level, but also on the practical one. Our senses that seem so perfect to us are designed at maximum savings to give us that much help that is necessary to move and operate in the environment without suffering damage.

For example, our hearing is capable of picking up very low sounds (about 30 Hertz of frequency) up to very high pitched ones, about 10-15,000 Hertz of frequency. However, there is an infinite variety of sounds, outside this range, that our ears cannot perceive. For example, ultrasound, commonly heard by many of our pets. Why do we only hear frequencies from 30 to 15,000 Hertz, and not higher or lower ones? Simple, because they would not be useful for our survival.

The lower audible frequencies can signal the roar of an approaching avalanche and the higher ones can signal the hissing of a snake. We don't need anything else to survive.

Forced by the limits of hearing, we place all the divinity of an art such as music within this range of frequencies, and we do not even imagine which symphonies could be composed (but above all heard) if the range were more extensive. For us all music is confined to the miserable narrowness of what our ears perceive, from 30 to 15,000 hertz. We do not even conceive of the existence of incredibly more elaborate and engaging harmonies. For us, there can be no other music than what our ears can hear.

We do not believe that there is a concert of the growing wood, a minuet of the blossoming flower, or the gloomy symphony of a rock that groans for years when it is about to detach itself from the mountainside.

To make up for these shortcomings, we add fake soundtracks to the nature films, suitable for our range of audibility. Obviously, these are not the real sounds.

And when they talk to us about shamans who are able to travel to these worlds and listen to these sounds, these hidden voices of nature, we don't believe it and we think those who tell us are crazy.

Our senses are not made to get closer to the universe but, much more prosaically, to prevent us from banging our toes when we walk or from putting hot food in our mouths when we eat or from getting into fights with some small creature too colorful or too thorny not to be poisonous.

Our senses are still those of Neanderthal Man, while we have now evolved well beyond the Sapiens sapiens stage. We are now Homo ciberneticus, but only culturally. We need to develop new senses, suitable for our new evolutionary stage.

Objectivity and subjectivity.

What we believe in determines the quality of our life.

We can choose between objectivity and subjectivity.

Objectivity necessarily entails our inability to intervene on the things that surround us. Things are like this, full stop, like it or not. From the objective point of view, everything is formed and happens externally to ourselves and involves us without the possibility of escape. We are mechanical toys that keep repeating the programmed task by marching and beating the tambourine both in the sun and in the rain, while the sun inevitably fades our colors and the rain rusts our gears.

On the contrary, subjectivity enables us to understand and accept things by mediating them, it is a web that we weave and through which we filter reality. Subjectivity provides powerful and effective tools to counter or support what life offers us.

Facing life from a subjective point of view means freeing oneself from the impositions of a massifying destiny and building a personal, mediated destiny, colored with different opportunities for each one.

A life lived subjectively is no longer measurable and weighable in the laboratory, because standard measures are no longer valid; each sets his own values which therefore escape the verification of external observers. They are no longer verifiable in the laboratory, but they are real for everyone in his heart, in his blood, in his belly, on his skin and above all in his soul.

Love, fidelity, generosity, friendship, trust in oneself and in others are all non-objective realities on which we model our life.

Therefore, there is a reality that goes far beyond the five senses, and allows us to move out of the dimensions of machines to become human beings, as we are.

The victory of subjectivity represents the defeat of materialism and the downsizing of the five senses which are no longer indicators of absolute truth, but simple useful aids to place our existence in the sensitive universe. In this universe, that of Newtonian physics, based on the three classical dimensions plus time, the five senses assist us in a truly exemplary way. However, when we consider the existence of psychic universes, free from matter, where different perceptions related to thought and soul prevail, then we realize that the five senses no longer help us, on the contrary, they often deceive us.

In recent centuries, materialism has entrusted the five senses with the task of absolutely representing all possible human experiences. In truth, materialistic science itself has found itself in difficulty in the face of what we now call phenomena related to matter, such as magnetism and electricity.

One of the most deceptive senses, when we want to use it for purposes other than those for which it was developed, that is, survival, is sight. Sight processes objects and living beings as forms, which are memorized to allow us to recognize them in subsequent encounters. This is not a precise memorization, because we are able to recognize a person's face even when his characters have changed a lot, after tens of years. Conversely, we remember with great difficulty what panorama was in the background or which cars and which people were passing on the road behind her. Yet, we have "seen" everything.

Psychology of the form.

Between the 19th and 20th centuries, experimental psychology devoted a lot of energy to investigating the mechanisms through which we relate to external reality.

The basis of the research normally started from the consideration that everything we perceive with the senses does not assume the same importance in the elaboration that we do. Listening to a conversation we normally "filter" it from all background noises. Observing a scene we focus on the essential fact while neglecting the outlines.

Figure 9 - The figure of Rubin reproduced above can have two distinct meanings, according to the subjective interpretation of the observer.

In other words, we are able to assign a hierarchy to the observed events, that is, to establish what interests us and what is not, in order to strengthen the understanding of only one part of the observed picture while neglecting the rest.

Our perception of what is happening around us is never the exact reproduction of reality. This means that our knowledge of the world never comes to be integral, because we do not faithfully record in the mind everything that happens, but only what interests us.

All perceived stimuli are immediately divided into "interesting" or not, so that most of them are discarded and lost forever.

Rubin's image, proposed in figure 9, can be interpreted in two ways. It is possible to see, in the center, the shape of a cup, or, on the sides, two human profiles. Through a quick observation our senses register one of the two meanings, but will it be the one that corresponds to reality? How many other "equivocal" frames do we perceive with the view from the reality that surrounds us?

The question is: who decides what interests us, and who decides what to discard and what, conversely, to keep?

With reference to these aspects of human perception, some researchers such as the Czech psychologist Max Wertheimer and the Germans Wolfgang Köhler and Kurt Koffka began a line of studies to which they gave the name of Psychology of Form (Gestaltpsychologie, or Gestalt psychology).

These studies were aimed at understanding the principles according to which some perceived elements come together to compose a "shape" that emerges in the background, ignoring other elements which, although present in the perception, are considered irrelevant, as if they were not themselves part of the perceived.

Koffka argued that there are things and "non-things": things are placed in a scenario of "non-things". The image of the Mona Lisa emerges against the backdrop of a landscape that is mostly of interest only to art critics. Instead, millions of people who have observed the painting, after a certain period vaguely remember the presence of a bridge (or is it a tower?). Usually no one remembers exactly how many cypresses are visible in the background.

Figure 10 - In the figure of Kanizsa represented above, a white triangle is clearly visible in the center. This, however, is not drawn in any way, and can only be imagined by considering the complexity of the design.

In reality, the habit of relating to the outside world through the five senses leads us to transform all of reality into things: it is not for nothing that when faced with an object or a fact that we do not understand we use to say in the same way: *What is this?*

To be accepted by our senses, things must have measures and weight, and must be placed appropriately in relation to the other things that surround them. A chair placed next to a coffee table is something that does not worry us, as it could happen if the chair appeared and disappeared or whirled in the air.

This is true as long as we observe the scene by analyzing it minutely in detail, that is, imposing on ourselves a complete analysis of the situation.

This is not normally the case. Our senses are much more hasty, and they must be since the scenes are like flowing rivers, they can change in the order of fractions of a second.

Therefore, the senses never completely analyze the whole situation but extract only the essential elements and this, in most cases, is deceptive as well as arbitrary.

A much cited example is that of the distribution of thirteen cards in a Scala 40 game. If a player received all thirteen cards of hearts, from the Ace to the King, he would cry out to the near-miracle. In fact, there is only one chance in 635 billion that a player will receive thirteen cards of the same suit. But if we think about it, any combination of thirteen cards can only happen once in 635 billion times, so any deal of cards is a small miracle.

The fact that a player receives thirteen cards of the same suit is significant and astonishing to us, but in reality it is an absolutely normal event. It is only we who attribute an extraordinary sense to a fact which, in itself, is absolutely ordinary.

For Gestalt psychology such an event is not significant because it is statistically rare, but because it stands out strongly against the background of thousands of other less linear card distributions.

Well-known examples are represented by the figures of Kanizsa.

Gaetano Kanizsa, born in Trieste in 1913 and died in the same city in 1993, was an Italian psychologist and painter. His father was a Hungarian Jew, his mother a Slovenian Catholic.

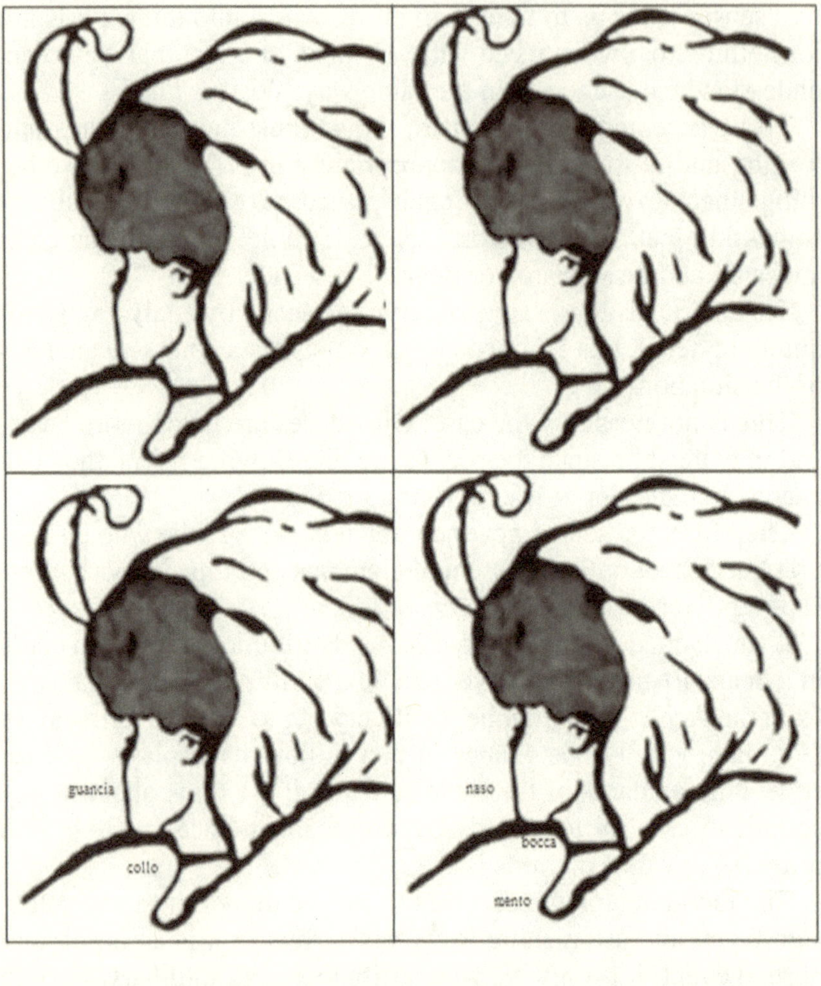

Figure 11 - This figure is more difficult to decipher, and shows the face of a young woman or, alternatively, the profile of an elderly lady.

After high school he enrolled in the philosophy faculty of the University of Padua, where Cesare Musatti introduced him to psychology. Struck by the racial laws, he was deprived of Italian citizenship. After the armistice of Cassibile he fled to Rome, where he was procured a position as a researcher at the Institute of Psychology of the CNR.

Figure 10 clearly shows a white triangle. This, however, is not drawn in any way, and can only be imagined by considering the complexity of the design. Although it is not there, it catches the eye much more than the other geometric figures. Not only that, its interior seems "whiter" than the white that contains the rest of the image. The image confirms that our senses are capable of making us see things that do not exist.

Figure 11 is more difficult to decipher, and shows the face of a young woman or, alternatively, the profile of an elderly lady. It is very difficult to grasp both meanings, if not knowing their existence. But our senses are not programmed to accurately analyze the constant bombardment of stimuli they receive. Therefore, the senses will register one of the two meanings of the figure, ignoring the existence of the other. But are we sure that they will register the one that most corresponds to reality? In 50% of cases, when faced with a representation of this type, our senses will deceive us. They will probably register the sense that is closest to what we would expect, that is, they will be "compliant" with our acquired cultural baggage. This means that, normally, any intellectual stimulus will be set aside in favor of the preservation and strengthening of the existing.

An assumption of Gestalt psychology is that our senses perceive and organize perceptions along some privileged, mysterious but solidly operating lines.

Koffka's studies have identified organizational principles of perceptions.

Among these the main ones are similarity, symmetry, proximity or distance, and others.

For example, a series of beeps in rapid sequence is interpreted as a danger signal; two shepherd dogs in a flock of sheep are identified as a distinct pair from the rest of the figures; a series of three black spots close together, plus three others close to each other but slightly distant from the first, is not interpreted as a set of six spots but as two sets of three.

A particularly interesting principle is that of temporal proximity. We tend to organize the perceptions and signals we receive from the outside also according to the temporal connection that unites them. Another important criterion is meaning. There are perceptions that leap to our eyes like beacons in the night, for how they are distributed over time and for the sense, the meaning they inspire us.

But here we enter another field, that of significant coincidences.

Coincidences, collective unconscious, synchronicity.

"What we know about us is only a part,
and perhaps a very small part of what we are without our knowledge. "
(Luigi Pirandello, Italian playwright, writer and poet).

More powerful than the most powerful computers.

In the previous chapter we saw how our senses scan an immense amount of data, which does not reach our consciousness as it is detected. The data undergoes processing according to procedures capable of extracting from the mass those of greatest interest to us. Subsequently, the interesting data are stored according to schemes with a high symbolic content. This allows us to recognize a face many years after it was last seen. In practice, we do not memorize the face as it is, but a shape of the face, some references that, after years, can be traced back to the original.

It is as if a particular software, residing in our consciousness, manipulates the naked and raw reality as it appears to the senses, transforming it into absolutely flexible forms or concepts, adaptable and reusable according to the needs. This allows you to categorize as a "chair" any object with one or more legs and a horizontal surface suitable for accommodating the lower back.

Odor recognition is an extremely powerful function. After decades, we are able to recognize one and immediately reconstruct the scene in which we first perceived it. For example, the smell of beached seaweed immediately brings us back to a school trip made by boys to a seaside town.

Face recognition is one of the privileged functions of our personal software. Whatever strange panorama appears before us, we can immediately identify the possible shape of a face in the arrangement of objects and elements. We easily recognize faces in the cloud formations, or in the rocky spurs or even in the damp spots present on the ancient walls.

In practice, every stimulus coming directly from the senses is processed by the software residing in our consciousness and is transformed in such a way as to make it usable in subsequent situations; it is therefore a learning and memorization software, whose main feature is being equipped with a powerful recovery algorithm.

In the early days of computer science, every student wishing to write small programs faced the problem of storing and retrieving data. For example, wanting to write a small program capable of indicating the capital of a nation, the student created a two-column file containing the nation-capital pairs (Italy - Rome, France - Paris, Spain - Madrid, and so on).

By supplying the computer with the name of the country (Spain), the machine scanned the entire file until it found the couple containing Spain and printed the result on the screen: Madrid.

This was so-called sequential access, and it could work as long as the data pairs did not exceed a few thousand, after which the system became too slow to be really useful.

Today we live in the age of search engines, capable of examining billions of data. Obviously, the search is no longer sequential. Search engines have highly sophisticated algorithms capable of extracting information in times of the order of milliseconds.

You can type "entanglement" in the Google search line, and you will see that in 0.36 seconds it will indicate the presence of about 9,700,000 results (search as of May 2018).

Well, long before man developed these algorithms, something similar if not more sophisticated was already present in the human mind. Our mental algorithms are capable of extracting information useful for our daily life in milliseconds by drawing from a mass of billions of billions of megabytes acquired over the years by our senses. But they will not just point out how many references exist: they will immediately select the most useful for us at that moment.

It is a wonderfully and astonishingly effective organization. Is it possible that this is nothing more than the fruit of various attempts, failures and new attempts at evolution, as the materialistic theses would like to support? Doubts are legitimate.

Does it only work with simple stimuli?

At this point we must ask ourselves if our software works only as we have seen it so far, processing without our knowledge the stimuli received from the senses, discarding or choosing according to mysterious criteria the interesting forms to keep and managing this amount of data without any intervention from our part. Or if we cannot hypothesize the possibility of our active involvement.

We have to start from an assumption. If our software is capable of discerning what can be useful and what cannot, it must necessarily be endowed not only with intelligence, but also with the capacity of discernment. That is, he must know what will be useful for us in the future, therefore he must have a vision of our existence that goes beyond the experience of the past and the present, but also extends into the future, in those circumstances and occasions that have not yet been realized, but they are precisely those for which he constructs the forms and archives them.

That is, the elaborations of our software will serve us in the future. Therefore, our software should know our future in order to process current information in a useful way.

If this is true, and it certainly is, then we can abandon the concept of software, because software is nothing more than a program built to perform certain actions. As intelligent as a software can be, it will never be more intelligent than the programmer who created it.

A software capable of knowing past and future goes beyond human capabilities. What operates in our consciences is not intelligent software, nor is it software developed by chance in the course of the evolutionary vicissitudes of the species. It is an Intelligence, or rather a pre-existing Wisdom, which we can define as Universal Intelligence.

Universal Intelligence knows our past, present and future because it exists in a different dimension from the one in which we are used to living, made of matter subject to the rules of Newtonian physics (gravity, time, and so on).

The level of Universal Intelligence must be purely psychic; it is a level made up of energy and information, in which matter is absent.

Is it possible that such a level exists? Certainly yes, as we will see in the following chapters. Quantum physics has revealed and experimentally confirmed the existence of this level.

Certainly quantum physics has not confirmed in any way the existence of a Universal Intelligence, but we like to imagine that there is, otherwise we could not explain in any way a whole series of phenomena, including the things already said about the intelligent storage of forms reworked from the raw perceptions of our senses.

The peculiarity of this Intelligence is its ability to establish a sort of interactivity with our consciences. In addition to the practically automatic role of constructing and memorizing the forms useful for our survival, the Intelligence constructs forms of life direction. Furthermore, the Universal Intelligence constructs proposals aimed at change, at correction, often simply at improving our relations with reality. These proposals are forms of communication that have been studied for decades by the Swiss psychologist Carl Jung He has called them *meaningful coincidences, or synchronicities.*

The principles of organization of our perceptions, managed by what we have called Universal Intelligence, do not apply only to reality perceived moment by moment, understood as a succession of distinct phenomena, but also to the perception of facts, phenomena and coincidences that could normally appear. devoid of any relationship or connection between them. Sometimes these perceptions. emerge in a completely spontaneous way, in our consciousness We see logical connections and intuitions emerge that give meaning to these coincidences and make them meaningful for our understanding.

We can say that Intelligence builds synchronicities for us and enables us to intuit the message that is hidden there. This is possible when we appreciate the possibility and the presence of synchronicities, avoiding considering simple fortuitous cases.

The clock that stops at the moment of the death of its owner constitutes a significant coincidence, due to the contemporaneity of the two events; this would not have been the case if the clock had stopped two days later or earlier.

Likewise, an encounter with an acquaintance from the past, after having dreamed it the previous night, is considered a significant coincidence.

Carl Gustav Jung

Carl Gustav Jung was one of the leading intellectual figures in psychological and psychoanalytic thought, psychiatrist, psychoanalyst, anthropologist and Swiss philosopher (fig. 12).

He was born on July 26, 1875 in Kesswil, Switzerland, to Paul Achilles Jung, a Protestant theologian and pastor, and to Emilie Preiswerk. He graduated in medicine and in 1900 he entered the psychiatric hospital in Zurich.

Initially he was close to the thought of Sigmund Freud, agreeing with his interpretations of psychic phenomena. The two met in Vienna in 1907, and on that occasion they talked for thirteen hours.

As a result of this collaboration Jung began to be called the "dolphin" of psychoanalysis, and it was expected that he would be Freud's successor at the head of the psychoanalytic movement.

Dissension and break with Sigmund Freud.

Instead, after only a couple of years there were the first signs of their disagreements, which would give rise to the two main historical orientations of psychoanalysis, understood by Freud as therapy and by Jung as a way to get to know the psyche.

This double vision quickly took hold, so much so that, as early as 1909, Clark University, having planned a cycle of lectures in the United States, invited both of them.

During the voyage by ship, the two scientists decided to practice analyzing each other's dreams. Jung, however, was not satisfied with Freud's collaboration because, in his opinion, he maintained an attitude of reticence, trying to hide some aspects of his private life that would instead have been indispensable for a better analysis.

When he expressed this perplexity Freud did not deny but affirmed that the reason for his reticence lay in the fact that he could not divulge aspects of his own personality that would jeopardize his authority. From this episode Jung's esteem for Freud began to fade and so did his appreciation for his theories.

To this was added the accentuation of a tormented itinerary of conceptual differentiation, which reached its apex when, in 1912, Jung decided to publish his work (Libido. Symbols and transformations) in which he exposed all his theoretical dissensions.

In this book he manifests his orientation, aimed at extending the analytic research from the story of the individual to the entirety of humanity: in fact, according to Jung, in addition to an individual unconscious there is a collective unconscious, which manifests its presence and acts through archetypes.

In this vision, the life of the individual becomes a path through which the realization of the personal self takes place. Jung calls this the discovery process.

After the publication of this book, his relations with Freud broke off. About Freud's thought he said: "The brain is seen as an appendage of the genitals."

Regarding the contents of the two interpretations, and their differentiation, we can consider that Jung initially shared with Freud the hypothesis relating to mental illness. According to these hypotheses, the understanding of diseases requires knowledge of the patient's individual history and of the "processes of repression" relating to contents of an exclusively sexual nature.

Later, however, Jung formulated the hypothesis that the phenomena of the psyche are not those described by Freud. According to Jung's new theory, those phenomena were manifestations of an energy present in nature, the *libido.*

Libido and symbols.

The libido is a kind of vital impulse, a drive that dynamically aims at the life and conservation of the species.

Therefore, in psychological therapy it is not necessary to keep in mind only the biological component. This would be a pessimistic approach. Conversely, it is also necessary to keep in mind the spiritual component. This component gives the individual a healthy tension aimed at achieving the improvement.

Figure 12 - Carl Gustav Jung. His thought postulates that, in addition to an "individual unconscious", there is a "collective unconscious" which expresses itself through archetypes. The life of the individual is seen as a path, called the "individuation process".

According to Jung, the libido drives can also be diverted to intangible concepts and ambitions and therefore can generate a spiritual evolution. Psychic illness occurs when this evolutionary process is blocked; under these conditions neuroses arise. The libido works by producing symbols. Thanks to this, primitive man was able to transfer the immediate impulses directing them towards creativity, thus initiating the transition from the animal instinctive to the cultural level.

The symbols of the libido have contents that come beyond the individual consciousness, and can help transform the nature of man.

Some of these symbols are universal, common and shared by all individuals. Jung calls them archetypes, a term derived from the Greek. The meaning of "archetype" can be made more understandable if we define it as "original images", "models", "virtual images", "possible representations".

The archetypes correspond to the experiences matured by humanity throughout its evolutionary history as it developed consciousness. They are the memory of humanity, and are enclosed in a space that Jung calls the "collective unconscious". This unconscious is collective because it is not individual but shared, accessible by all humanity and by every single human being.

The archetypes residing in the collective unconscious have the power to intervene in the life of individuals by generating episodes that Jung calls synchronicity.

In the construction of this theory Carl Jung had the cultural and moral support of the famous Nobel laureate physicist Wolfgang Pauli, (Fig. 13) who encouraged him to explain his conclusions despite Jung hesitating for years to do so. Pauli persuaded him to publish together, in 1952, a book that contained it in the form of an essay entitled Synchronicity as a Principle of Acasual Connections

The symbol.

The symbolic character of the archetypes produces the consequence that even the synchronicities generated by them have symbolic aspects.

Figure 13 - Wolfgang Pauli, a very young student at the Döblinger Gymnasium in Vienna. He met Jung for a therapeutic treatment that soon resulted in a solid friendship and a quarter-century exchange of letters..

This is supported by experience, and some Jungian analysts do not hesitate to define synchronicities as "daydreams".

In fact, dreams are absolutely independent of our will and we could hardly intervene to drive them in our ordinary life. Likewise, synchronicities are unique events that arise in our lives arousing bewilderment and wonder. The sncronicitò are always surrounded by an aura of mystery due to their symbolic form.

The strange coincidences of which we are sometimes protagonists also have a symbolic value, but their repetition can guide us towards deciphering the symbol and understanding the message. When this occurs, it follows mysterious procedures. Sometimes it happens: our consciousness simply unfolds the meaning of the symbol before our eyes, although we cannot reconstruct the procedure according to which this could have happened.

According to the common meaning, the symbol is something that stands in the place of something else. Most of the time the hidden meaning of the symbol escapes us. As much as we seem to have it "on the tip of the tongue", it buzzes around us without us being able to grasp it.

In this sense, the symbol generates a tension between what appears and what escapes, it contains a part of the known and a part of the unknown. It represents a challenge for the conscience that suffers in the restlessness of an unsolved problem, an unanswered question, a pawn that has no place.

It is not impossible to win the challenge. In fact, synchronicities have the function of making the messages coming from the collective unconscious conscious through our individual unconscious.

A mysterious message flows from the collective unconscious, written in a language and with symbols built on the evolution experienced by all of humanity and therefore indecipherable to the maximum degree for the vision allowed by our extremely limited experience. Let's say that our conscience does not have eyes large enough to be able to visualize and understand the total meaning of the message.

When it passes through our individual unconscious, the message tries to make itself understandable to our conscience.

Mediated by the individual unconscious, the symbol dresses up with things that have happened that we no longer remember, or that we want to forget; rummaging through a warehouse of ruins and fragments of destroyed buildings that, however, were familiar to us at a certain time, and uses them to express himself more closely to our experience. While retaining its crypticity it becomes more decipherable.

If it were not connected to the collective unconscious, our individual unconscious would be just a sad useless thing, a cemetery of buried experiences.

Thanks to the collective unconscious, our individual unconscious becomes a precious tool, and is transformed into the medium of exciting life moments, of daydreams that can guide us towards a spiritual evolution and a better life.

Archetypes, ideas, concepts.

The word archetype comes from ancient Greek and can have the meaning of original, model, brand. According to some, it can also mean "beginning". It was first used by Philo of Alexandria, an ancient Greek philosopher also known as Philo the Jew, who lived in Alexandria in Egypt and was a contemporary of Jesus. Later the term was taken up by other Greek philosophers such as Dionysius of Halicarnassus and Luciano of Samosata.

The development of the theme of archetypes can be inserted, however, in the Platonic tradition, where they appear as ideas, that is, original models of sensible things. Some Neo-Platonists such as Plotinus regarded them as pre-existing ideas in the divine mind and models of created things.

Within this tradition, Jung considers archetypes as primordial images that constitute the collective unconscious, or even primitive forms on which myths and religions are based.

In analytical psychology the archetypes acquire an ethical and affective value for the subject, and condition him more or less consciously throughout his existence.

Indeed, the archetypes, as conceived by Jung, have been compared by many authors to mythical structures and folk tales. In fact, they would suggest many similarities of meaning with fairy tales and myths. Reciprocally, there are fundamental reasons in myth and religions that are regarded as archetypes. In his book Psychological Types Jung states: "The main mythological motifs of every time and of every race are probably of this type (archetypal)."

The fourth excluded.

Jung was a skilled lecturer, managing in any case to interest the public even when it was not made up of people introduced into the scientific environment. Contrary to Freudian prose, unadorned and measured, Jung knows how to express his rich and cultured personality even in the most technical or systematic works. In this chapter I mainly refer to three essays by him, presented in the form of lectures. The first, "The archetypes of the collective unconscious", was written in 1934 and revised in 1954. The second, "The concept of the collective unconscious", was written in 1936.

The third work, Synchronicity as a principle of acasual connections, published in 1952, is almost a compendium of his first works and is the one in which he fully expresses his theory on synchronicity.

Jung was very reluctant to publish this essay, and only in 1952 did his friend and correspondent Wolfgang Pauli persuade him to do so.

The two had collaborated for more than twenty years through an intense exchange of ideas, precisely on these issues.

In fact, despite being one of the founding fathers of quantum physics, capable of winning a Nobel Prize in 1945, Pauli very much shared Jung's ideas and would have liked the scientific environment to be able to overcome mistrust, skepticism and derision by considering the possibility of a parallel path and a common language between science and psyche, which could have allowed greater understanding.

The intellectual confrontation between Jung and Pauli generated research known as "the excluded quarter". In fact, the object of their studies tended to resolve many of the doubts raised by quantum physics regarding the completeness of the triad model of classical physics: time, space, causality; Jung and Pauli imagined an "excluded quarter" which was given the name of synchronicity.

Pauli was fascinated by the possibility of adding the fourth element to the triad. Jung, for his part, was fascinated by early studies of subatomic particles that began to reveal incredible properties, such as those of mutating into waves, or disappearing and appearing in other remote points, or communicating over immense distances through mysterious non-physical connections. He saw nascent quantum physics, and even Einstein's theory of relativity, as possible scientific confirmations of his psychological concepts of synchronicity and the paranormal.

Pauli possessed a profound intuition (which, according to him, derived a lot from his dreams) and an exceptional ability to think.

At a very young age he published an article on the theory of relativity which was published by the Enzyklopadie der mathematischen Wissenschaften. When Einstein, the father of the theory, read it, he commented:

> "Anyone who studies this work would not believe that it was written by a young man of only twenty-one. I don't know what to admire more: the psychological understanding of the evolution of ideas, the accuracy of mathematical deductions, the profound intuition, the ability to present the work with systematic lucidity, factual completeness or critical infallibility."

Unfortunately, Jung, like all geniuses, had always clashed with the coldness and misunderstanding of his most original ideas on the part of his colleagues and his working environment. His regret, and perhaps his suffering in not being able, from a young age, to involve even the closest people in his intuitions, is evident from the pages of Memories, dreams, reflections, his diary-confession entrusted to the pen of Aniela Jaffè. Therefore, in his maturation, he had instinctively assumed an attitude reluctant to make public his theories of his most daring.

It was Pauli who overcame his reticence by convincing him to publish a book together, in which two essays appeared; one by Pauli himself entitled The influence of archetypes on Kepler's scientific theories and the second, signed by Jung, Synchronicity as a principle of acasual connections.

Archetypes as primordial images.

The book published with Pauli represented the synthesis of decades of work. The first time that Jung used the term archetype dates back to 1919, on the occasion of a symposium held in London.

> "In the unconscious we also find characteristics that have not been acquired individually, but inherited, that is, instincts understood as impulses for activities that come, without conscious motivation, from an inner constraint. To these are added the existing a priori, that is congenital, forms of intuition, that is, the archetypes of perception and understanding, (...) As the instincts induce man to a specifically human behavior, so the archetypes force perception and the intuition to specifically human formations.
>
> The instincts and archetypes of intuition form the collective unconscious".

Jung soon developed the conviction that archetypes were ideas external to the individual consciousness, which had always existed and were closely connected to primordial myths. The event that led him to convincingly embrace the thesis of the relationship with myths is told by himself in the essay "The concept of the collective unconscious" written in 1936.

"Around 1906 I came across a very strange delusional illusion, in a paranoid schizophrenic who had been interned for many years. The patient suffered from a young age and was incurable. He had studied at a state school and had been hired as an office clerk. He had no particular inclination and I myself knew nothing of mythology or archeology at that time. One day I found the patient by the window, shaking his head and blinking at the sun. He told me to do the same thing, because that way I would see something very interesting. When I asked him what he saw, he was amazed that I could not see anything and said, "Surely you see the sun's penis... when I move my head back and forth, it moves too, and that's where the wind comes from. " Of course I didn't understand this strange idea at all, but I took note of it.

About four years later, then, during my mythological studies, I came across a book by the late Albrecht Dieterich, the well-known philologist, who shed light on this fantasy. The work, published in 1910, deals with a Greek papyrus from the National Library of Paris. Dieterich believed that he had discovered a Mithriac ritual in one part of the text. The text is undoubtedly a religious prescription concerning the implementation of certain spells in which Mithra is named. (...) In the text we read the following indications:

"Breathe in from the rays, breathe in three times as much as you can and you will feel yourself being lifted and walking upwards and you will feel like you are in the middle of the aerial region ... You will see the path of the visible gods through the disc of the sun, who is God my father. So too you will see the so-called tube, the

> origin of the helpful wind. Because you will see something like a tube hanging from the disc of the sun. You will see this towards the western regions, as infinite as the wind blowing from the east. If the other wind is directed towards the eastern regions, you will see the reversal of vision in that direction "
>
> (...) Now, it is absolutely impossible that the patient could have had any knowledge of a Greek papyrus published four years later".

Jung considered this episode as a first evidence of a close relationship between the archetypes that materialize in the human mind and the symbols of ancient mythologies.

The collective unconscious.

The collective unconscious is a concept of Jung's analytical psychology. In opposition to the personal unconscious which is individual, the collective unconscious is shared by all men and derives from the common ancestors of all humanity. In this sense it represents a universal psychic container, that is, a part of the unconscious common and shared by all of humanity. It contains the archetypes, that is, the forms and symbols typical of all peoples of all cultures (fig. 14).

It is important to note that archetypes do not derive from the individual's experience but exist a priori, as a heritage of the experience of all humanity. Some use to divide the collective unconscious according to different stratifications: the deepest derives from the archaic roots of the past; the average stratification contains the values and social customs of the moment that is being lived, and a third stratification contains the potential and possible future evolutions of humanity.

Jung presents and defends his theory in the essay The concept of the collective unconscious, published in 1936.

He introduces the concept by stating that the personal unconscious owes its existence to personal experience and is essentially composed of contents that have been conscious, even if they may then have been forgotten or removed.

Conversely, the contents of the collective unconscious have never been in the individual consciousness and therefore have never been acquired individually, but owe their existence exclusively to inheritance.

Consequently, he differentiates the two types of unconscious by referring to their content: the individual unconscious contains above all complexes, the collective unconscious, on the other hand, is essentially formed by archetypes.

Jung defends his idea of the archetype as a pre-existing form, citing other researchers such as Lévy-Bruhl who in his studies on the psychology of primitives defines archetypes as "collective representations". Jung also cites Hubert and Mauss, who in their studies of comparative religion define the archetypes "categories of the imagination". He also cites Adolf Bastian who in his ethnological studies defines the archetypes as "elementary thoughts" or "primordial thoughts".

Strengthened by these scientific convergences, Jung exposes his thesis, according to which:

> "In addition to our immediate consciousness, which is entirely personal in nature and which we believe to be the only empirical psyche (even if we add the personal unconscious as an appendix) there is a second psychic system of a collective, universal and impersonal nature, which it is identical in all individuals. This collective unconscious does not develop individually, but is inherited. It consists of pre-existing forms, the archetypes, which can become conscious only at a later time and give a specific form to certain psychic contents".

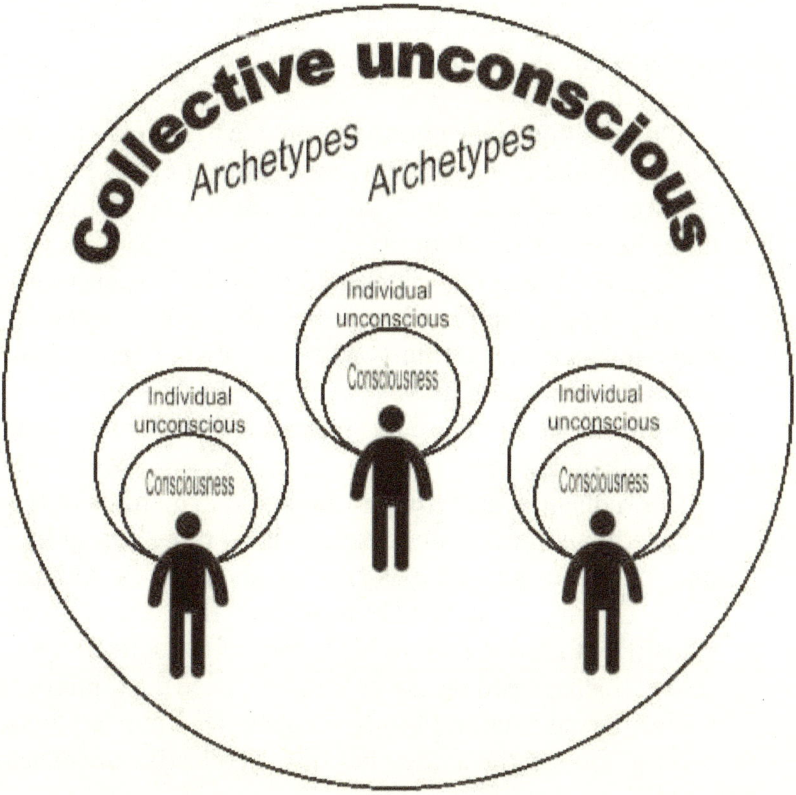

Figure 14 - The levels of consciousness according to Carl Jung.
1. The collective subconscious is common to all of humanity
2 Archetypes come from the collective subconscious.
3 -4 Each individual has an unconscious and an individual consciousness.

Jung then goes on to illustrate the meaning of the collective unconscious and begins by referring to therapeutic psychology in the vision of Sigmund Freud and Alfred Adler, which is exclusively personal in nature and denies any external contribution or connection to the individual..

Jung highlights the contradiction due to the fact that, however individual it may be, this view also refers to factors that are not merely individual, such as the sexual instinct or the drive for self-affirmation.

Not even Freud or Adler could deny the existence of instincts existing a priori and present both in humans and in animals. Instincts are certainly not personal but universal and inherited components of the psyche, and often operate above consciousness. In this sense, instincts show very close analogies with archetypes.

Consequently, according to Jung, the hypothesis of the existence of a collective unconscious is no more daring than the one that advocates the existence of instincts. On the basis of this reasoning, he defends himself from the accusation of mysticism that is addressed to his theory of the collective unconscious. Just as instincts are able to influence the thinking, consciousness and action of individuals in a completely mystical way, so too can the archetypes of the collective unconscious.

The identification process.

In the first chapter of his work Aspects of the contemporary drama Jung writes as follows:

> "Archetypes are like river beds abandoned by water, which however can welcome it again after a certain time. An archetype is similar to a mountain gorge into which the current of life has long

> poured: the more it has dug this bed, the more it has preserved this direction, the more likely it is that, sooner or later, it will return.."

The psyche is essentially composed of a conscious part (the conscience) and an unconscious part, in which we can distinguish the individual unconscious and the collective unconscious.

The dynamic between the two parts (conscious - unconscious) is considered by Jung to be the engine that allows the individual to face a long journey aimed at realizing his own personality.

The interaction between the conscious and the unconscious generates a process he calls "individuation". In this path the individual encounters and confronts himself with unconscious aspects (archetypes) of his own personality: by facing them he can broaden his conscience.

The process of individuation represents a path of integration and unification of the complexes that make up the personality, that is, a progressive approach of the Ego with the Self, through a path that leads to the consciousness of the Self and to the liberation of the Ego from inner conflicts.

Jung believes that the task of psychoanalysis is to establish a hierarchical relationship between "Self" and "I", considering precisely that the former is the whole while the latter is only a part. This relationship should allow a resumption of the harmonious rapprochement between the two, when this had been arrested by a neurosis resulting from a stiffening of the instances of the "I" with respect to the individual needs of the Self (hence the name: individuation process).

The approach takes place through the deciphering of the meaning of the symbols that the individual encounters during his life. Symbols can be found both in the internal world, through the regressive and progressive movements of the libido, and in the external world, thanks to the episodes of synchronicity. Therefore, synchronicities are a valuable aid in the development of the individuation process.

In this light, individuation is the fruit of the continuous collaboration between Consciousness, Unconscious, I and Self. The individuation process represents a spiritual journey towards greater self-awareness. Progress in the journey is recognizable by its stages.

First stop: the archetype of the Shadow.

The first stage is characterized by the archetype of the Shadow, i.e. all those aspects that the individual does not know or wants to hide about himself, as well as everything that has been removed following the compromises due to the education and influences of the environment.

The Shadow represents the most unpleasant and negative part of the psyche; it coincides with even the lowest instinctual impulses, which one usually tends to repress. It also represents the vices and inclinations that the individual tends to deny being of him, but which, despite this, influence his behavior and manifest themselves with detestable sides of the character or with "temptations" that the conscious part tries to ignore.

This archetype can manifest itself in dreams and symbols with demonic figures, monsters and chases, nightmares lived in the dark.

An archetype often cited together with the Shadow is that of Persona.

The Person (from the Latin, "actor's mask") is an archetype that indicates the role that the subject plays in the social context in which he acts. Person is precisely the mask, the covering identity under which we act, and thanks to which we are what others would like us to be and what we believe we are.

Second stage: the archetype of the Anima or the Animus.

The second stage is characterized by the encounter with the archetype of the Anima (for the male) and the Animus (for the female). Despite the name that recalls the soul, it does not refer to any concept of a religious or dogmatic type.

The image of the Soul is projected by men onto women, while in women it is the corresponding archetype, the Animus, that is projected onto men.

The Soul represents the contact with the psychic and mental aspects, that is the first approach of an individual with his psychology, almost an initiation rite. It is the door that allows access to the transcendent world.

It is an archetype of great communication because it presents many unconscious images, creates misunderstandings, constructs illusions and consequently also crises.

In this sense, the confrontation with the soul is much more demanding than that with the Shadow; it involves entering territories where there is a risk of being crushed. Everything that the Soul touches becomes numinous, that is, absolute, dangerous, subject to taboo, magical.

The Animus (male component of women) represents typically practical and concrete, rational and real aspects. He is represented in dreams with war, the blacksmith and similar symbols. It has contents of reflexivity, control, analysis, weighting, rationality, calculation, decision, planning, distinction.

These two archetypes give us a broad view of how dual the psyche can be.

According to Jung, each archetype contains an aspect of life and its opposite, it being understood that both have their own value and their place. On the contrary, Freud's thinking is centered on conflictuality and the struggle for the elimination of one of the two.

According to Jung, the duality of the psyche involves submission that overcomes abuse and love that overcomes hatred. A harmonious collaboration must also exist between the conscious and the unconscious.

The Jungian duality also remains valid with respect to the diversity of genders, as the psyche has within it masculine and feminine energies, and every man has a feminine side, just as every woman has a masculine side. It is obvious that each one has a dominant energy, but despite this it hosts, albeit in the background, also the opposite one.

In the bas-reliefs of Indian temples, the masculine and the feminine are represented as a man and a woman embraced; allegorically they are two universal figures. The initial unity of the two genres also transpires from the religious or mythical narratives, where we read that Eve is born from a rib of Adam and Athena comes out of the head (or calf) of Zeus.

Third stage: the archetype of the "Old Sage".

The third stage corresponds to the encounter with the archetype of the "Old Sage", who can be understood as the paternal male figure, mirroring the archetype of the Grance Madre. The Old Sage contains all the experience of the individual, and with it all his ability to reason and foresee. It represents the transformative goal that one is reaching after having confronted oneself with the archetypes of the previous phases. It is the condition in which one becomes capable of evaluating the past, present and future. The Wise Old Man is portrayed as a philosopher capable of advising.

The "Great Mother" represents the final goal of female psychology. In Jungian psychology she is an archetype of great and ambivalent power, at the same time destroyer and savior, who nourishes and devours. She is clothed with sacredness, such as to be one of the numinous powers of the unconscious.

Fourth stage: the archetype of the Self.

The archetype of the Self is the final motive of the individuation process, that is the reason for existing of the individual, which reveals itself to him and opens up in front of him like a blossoming flower.

It is represented as light, quatern, center of things. Its symbol may be the mandala, a figure that in the Buddhist and Hindu religious tradition represents the cosmos and is painted on fabric or frescoed on the walls of temples.

The Self represents the individual himself, with all the experience and knowledge that he has accumulated in his path of individuation. When the individual encounters the Self, it means that the neuroses are defeated and inner harmony has been achieved.

Therefore, the Self is the goal, the culmination of the realization of the personality, a peak in which all the conscious and unconscious aspects of the individual converge and unify.

Other classical archetypes represent universal images that simultaneously express positivity or negativity: Water, the Number, the Hero, the Wizard, the Rebel, the Sovereign, the Innocent, the Apollo, numerous figures of religion, etc.

When the archetypes refer to mythical, religious or legendary figures, they are called by Jung "numinous powers", taking up a term already coined by Rudolf Otto, German theologian and historian of religions.

Jung and alchemy: individuation and Opus alchemicum.

The last book written by Jung in 1956 is called Mysterium Coniunctionis and deals with the archetype of the same name, which refers to the sacred or "alchemical" marriage between the Sun and the Moon. Jung focused on alchemy in many of his works, trying to highlight the significance of Opus alchemicum (Work of alchemy) as a spiritual quest.

His research aimed at discovering the analogies existing between alchemical processes and those related to the psychic sphere. In particular, he analyzed the phases through which the alchemical work would take place, that is, the procedure for obtaining the philosopher's stone. This process involves some stages, which originally were four:

- Nigredo or blackening, associated with the earth element.
- Albedo or bleaching, associated with the element of water.
- Citrinitas or yellowing, associated with the element of air.
- Rubedo or redness, associated with the fire element.

These four stages would correspond to the psychic process of individuation.

In particular, the first stage of identification, that is, the comparison with the archetype of the Shadow, would find correspondence in the phase of the alchemical process called nigredo.

Nigredo is a Latin term which means "black color" or "blackness". It marks the Black Phase of Opus, the beginning of the path of creation of the philosopher's stone. It is the stage of putrefaction and decay. This is the most delicate moment, symbolized by a black crow. In this phase all the ingredients must be left to die, macerating and cooking them for a long time until they form a dark mass.

The second stage, that is the meeting with the Anima or Animus, would find correspondence in the Opus phase of the albedo, a Latin term which means “whiteness”. This phase is symbolized by a white swan and is called Opera al Bianco. It consists in the purification of the shapeless mass resulting from the nigredo, to wash away the impurities and prepare it for the next phase. In chemical language it corresponds to distillation. Metaphorically, with this passage the soul is freed from the bonds of carnality.

The third stage, that is the meeting with the Old Sage, would find correspondence in the phase of the alchemical process called Citrinitas.

In truth, this phase was often considered as part of the albedo, while other traditions prefer to consider it as a phase in its own right.

Citrinitas is the phase of yellowing or burning of the matter of Opus. Its color is citrine yellow, a symbol of will and a dynamic conscience. From the initial putrefaction a distilled compound was obtained, destined for sublimation.

The fourth stage, of the individuation that is the encounter with the archetype of the Self, would find correspondence in the phase of Opus called Rubedo.

The Latin term rubedo can be translated as "redness". It represents the Red Phase and is the completion of the chemical transmutation, which should coincide with the realization of the philosopher's stone capable of converting base metals into gold. Rubedo occurs due to the effect of fire, that is, of the Spirit. Its symbols are the phoenix, the red rose, the egg, a crowned King.

Rubedo also has other symbolic values, linked to the fact that red in the alchemical tradition represents the intermediate color between black and white. Therefore, it represents the conjunction of opposites and the union between spirit and matter or between masculine and feminine.

Finally, just as the nigredo corresponded to the physical body of the alchemist, and the albedo to his soul, the rubedo corresponds to his spirit, that is, to the highest part of the three constituents of the human being.

Synchronicity.

It is in the region of the heart that spontaneous recognition of connections takes place, which the mind often understands only later on. These connections were at first almost inexplicable but perceived with absolute certainty they are defined as synchronicity,
or coincidences, or simply miracles.
(Ruediger Schache, Advice from the Heart)

Jung fully exposes his theory on synchronicity in the essay "Synchronicity as a Principle of Acausal Connections". He begins his reasoning by defining the concepts of causal - acausal.

Jung begins his exposition by emphasizing how the concept behind the laws of nature is that every event is due to a cause. However, this is not always true; there are events in nature that are not linked to each other by cause-effect relationships and therefore must be interpreted differently.

It would be a futile undertaking to look for these events in the macrophysical world, if only because it would be impossible to give them an explanation. Let's imagine a billiard ball moving without any push. The ball has not received the impact of another ball, the thrust of the wind, it is not on an inclined plane. The fact would not be credible. not even if many witnesses told it.

It is a confirmation of the concept of determinism, according to which every physical phenomenon or event is necessarily determined by another phenomenon or event that happened previously.

In order for a phenomenon to be taken into consideration, a single testimony is not enough, a great many are needed: despite this, the phenomenon will continue to be denied in the absence of repeatability, that is, the possibility of reproducing it normally. Repeatability must be possible under any similar conditions.

Sometimes, however, even a rare phenomenon can be taken into consideration, as happens in biology, where a single specimen of a species makes the existence of the entire species credible.

The question that Jung asks himself is: is there, in an absolutely general way, a possibility of obtaining confirmation of uncaused events? Is there a series of unjustified events? His answer is positive. Jung refers to the world of "chance", in which events do not seem linked by causal relationships with other events.

When we talk about the case it is customary to argue that even the events defined as "randomness" or "coincidences" can be explained. When these events cannot be attributed to a cause, it only depends on the fact that it is not known.

But, Jung continues, the principle according to which every event must have a cause is relative, that is, it is applicable only to the context of the macrophysical reality, the one in which we live.

Consequently, while admitting that the vast majority of events must have a cause, this does not exclude that some events may be Among the vast majority of events related to a cause, there are the rare cases where no cause exists. These cases can be traced to just where a causal relationship seems unthinkable. Jung gives an example:

> "If I am forced to find out that my tram ticket has the same number as the theater ticket I buy immediately after, and if during the same evening I still receive a phone call in which that same number is listed as my phone number, a causal relationship seems to me very unlikely, and even the most daring fantasy would not allow me to think that a relationship can exist, although each case taken by itself just as evidently has its cause.

It is obvious that all coincidences could be like targets hit by chance, but this theory can hold up as long as it does not exceed a reasonable level of probability. When this threshold is exceeded it is evident that the complex of coincidences becomes inexplicable and cannot be traced back to any cause.

Jung admits the possibility that, however, a clustering of coincidences may be due solely to chance.

However, there may be groups of coincidences that take on a different and highly significant value for whoever is the protagonist. These are the cases in which the subject involved manages to connect the coincidences, giving them meaning.

The occurrence of these groups of coincidences contains a numinous, that is, mysterious, sacred component that makes them significant, and when this happens Jung gives the group the name of synchronicity.

Illustrating significant coincidences, Jung groups them into three categories.

The first foresees that there is a link between a psychic state of the subject and a contemporary and objective event, which corresponds to the psychic state. An example is the episode narrated by

Jung himself: while a patient told him she had dreamed of a beetle, an insect of that type crashed into the window of the study.

The second category of significant coincidences foresees that there is a link between the psychic state of a subject and an external event, which occurs in another place but can be verified later. The countless reports of dreams announcing the death of a person, which turns out to have occurred at the same time as the dream, can be used as examples.

The third category provides that there is a link between the psychic state of the subject and an event that has not yet occurred, but which will happen in the future. For example, you may suddenly think about someone you have lost sight of and meet them the next day.

In all three cases the need for the existence of a psychic state is evident that allows the subject to intercept the signals that announce the event. The subject gives a meaning to the events and confirms in himself that charge of expectation that will coincide with the completion of the event.

Unfortunately, the signs that we could define premonitory are often ignored, however numerous they may be, because of course we are used to seeing the world according to the parameters of deterministic macrophysics and to absently considering unusual events and strange coincidences, downgrading them to unimportant cases.

Birth of a synchronicity.

In his writing Synchronicity as a Principle of Acausal Connections published in 1952, Jung cites an example of synchronicity:

> "To cite one example among many, I have noted the following case dated 1 April 1949. Today

> is Friday. We have fish for lunch. Everyone remembers en passant the use of "April Fool". During the morning I had written down an inscription: “Est homo totus medius piscis ab imo”. In the afternoon, a former patient whom I had not seen for months shows me some particularly suggestive pictures of fish, which she has painted in the meantime. In the evening they show me an embroidery representing sea monsters in the form of fish. On April 2, early in the morning, a former patient whom I had not seen for several years tells me a dream in which, finding herself on the shore of a lake, she sees a large fish swimming decisively in its turn and "lands", so to speak, at his feet. In this period I am busy with research on the historical symbol of the fish. Only one of the people mentioned above knows. There is an obvious suspicion that in this case it may be a significant coincidence, a casual relationship. I must admit that this concentration impressed me. For me it had a certain numinous character. In such cases it is notoriously said: "It cannot be pure chance!". Fish often emerge as symbols of unconscious content..”

In the language of the psychology of form, significance is the quality that distinguishes things from non-things, the central scene from the background, the relevant actions that compose non-relevant shapes and outlines.

We could divide perceptions into photographs and films, to suggest that there are not only instant perceptions, comparable to photos. We often perceive some more structured situations, with their own timing, complex events that stand out against the background of other events, as happens in films in which the central action captures all the attention at the expense of the outline.

In this case we are talking about events, circumstances, significant coincidences, each with its own significance.

Figure 15 - Let's imagine that the dotted square encloses a period of a person's life, which we refer to as Philip. A day, a week or more. Each point represents one of the countless events that occurred during this period.

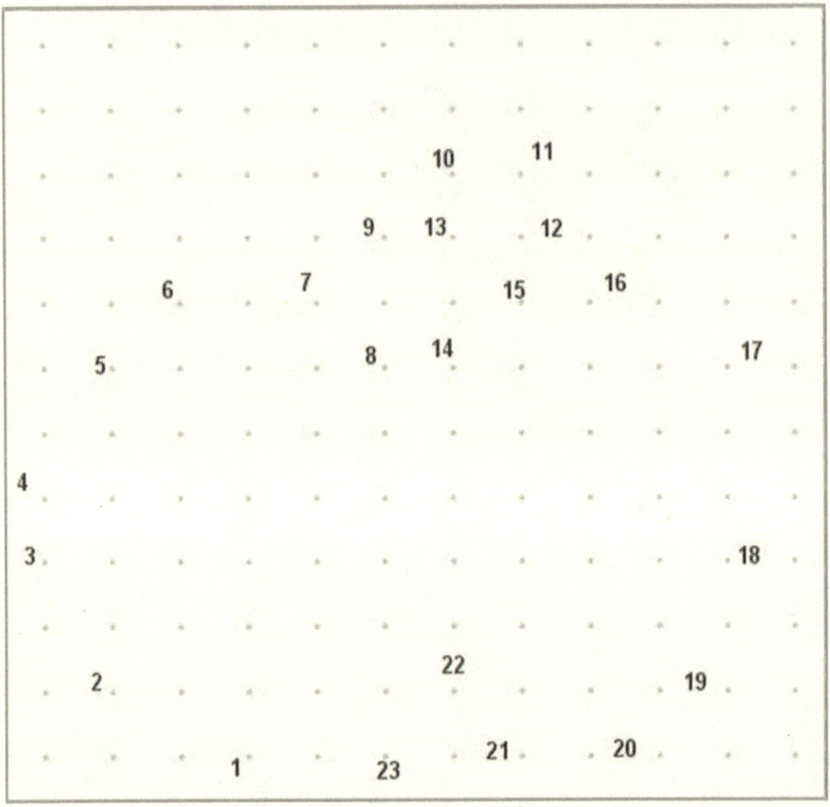

Figure 16 - Of course, some episodes are more relevant than others. Then there are some (those highlighted in bold) that take on a particularly significant value for Philip, even if he himself cannot explain the reasons.

All together they contribute to building a more complex form, made up of film clips, each independent of the other, but overall endowed with a symbolic nature relating to the events and needs of our life. In this case we are talking about synchronicity.

Probably the simplest method to explain what a synchronicity is is to simulate the process by which it is built, that is, it takes shape in the consciousness of an individual.

We will therefore imagine to reconstruct the passages through which a synchronicity could emerge and become effective in the conscience of a friend, to whom we assign the conventional name of Filippo..

Philip and his marriage.

Figure 15 symbolically encloses in a delimited area (a square) a period of time in Philip's life, reasonably one week. Each dot enclosed in the area represents one of the countless events that involved him in that period, from tying his shoes in the morning to turning off the light and going to bed at night. Each dot has its importance, but surely some events are more relevant than others, such as for example it could be a praise received at work compared to throwing a waste in the bin.

However, there are some situations that remain particularly etched in Philip's mind, even if he cannot explain why. These are events that are not connected to each other nor have any connection with the ordinary course of his day.

They are simply like flashes that dazzle the mind as they could dazzle the eye. As you know, the effect of a flash takes some time to free the eye from discomfort, and so too the strange sensation aroused by some events continues to flicker in the mind, for no particular reason.

In figures 16 and 17 I wanted to represent everything as if it were the famous game “Connect the dots”.

Indeed, Philip must strive to connect strange facts, dreams and coincidences with each other to make sense of them, that is, to build a shape that gives a common meaning to events that apparently nothing in common should have.

The night between Monday and Tuesday Filippo has a strange dream. He is on an island in the Pacific Ocean, and is involved in a party organized by the villagers. Many dance, others play instruments characteristic of the place and still others offer wreaths of flowers, all in an atmosphere of serene joy. A young woman approaches Philip and puts a crown of bright and fragrant flowers on his head, which rest on her shoulders and descend on his chest. Suddenly, while he participates in the festive atmosphere, Philip lowers his gaze and sees a small insect that slips out of the chalice of a flower and begins to nibble on its petals. Soon after, other insects appear and attack all the flowers in the crown, which quickly wither and rot. The dream is turning into a nightmare; Philip wakes up with a start and realizes that he has only had an unpleasant dream experience, but the image of the beautiful flowers that rot under his eyes remains vivid in his consciousness.

The following day he discovers that a documentary on the military organization of the Roman Empire is being broadcast on a TV channel he does not usually watch. He has always been interested in the history of ancient Rome, so he decides to follow the show. The film presents the organization of a "castrum". The castrum was the camp that each legion had to build in a few hours at the end of a day's march. At first, the military dug a moat and built a palisade inside it. Inside the palisade The land was divided by the "Via Principalis" and the "Via Praetoria". The tents of the command, officers and soldiers were placed in the various sectors. At the end of the construction the commander offered a sacrifice to the gods to protect the newly erected camp.

"How strange, - Philip thinks, - they offered sacrifices on every occasion, so why in such a well-organized camp did they not also erect a small temple for the gods?".

This thought surprises him: he would never have imagined worrying about building a religious temple. Filippo wants to put the thought aside but this insistently turns around in his head.

The third curious coincidence occurs after a couple of days, while he is working in the office, and he finds himself grappling with a felt-tip pen which, having been left uncapped for too long, he does not want to write. Impatient, he throws it in the trash. Guido, a colleague, intervenes and tells him:

"Don't throw it away, give it to me. With a little patience I will have it written again and I will use it for the announcements to put on the notice board of the club ".

Guido, in fact, is secretary and factotum at the club of bowls that both attend and where, due to the scarce resources, he tends to save on everything.

Of course, Filippo has nothing against it, but he is struck by the image of the notice board of the club where we are reduced to writing with salvaged markers.

In a few days, incomprehensibly, Filippo was attracted to a series of curious and completely disconnected facts. He begins to wonder if this all makes sense. Fresh flowers donated in a feast that soon after are ruined by parasites, a pagan temple that worries him for the lack of him, a notice board on which he writes with out of stock markers.

Suddenly, as in a flash, the common link literally explodes in Philip's mind.

The common denominator is a church, of course. Not just any one, but the little church where he got married. The flowers are those of the ceremony, the temple is the church itself, the bulletin board is that of the marriage banns.

In our "Connect the Dots" metaphor for the game, as seen in figure 17, Philip connects the dots and obtains the shape of the church, which he immediately identifies as the one in which he got married.

Still not enough, the church is simply a symbol, the sum of many symbols.

PPhilip must come to understand the message that is hidden behind these symbols: the flowers that are attacked by insects and rot, the votive temple that is not there, the out of place marker. All this connected to the church where his marriage adventure began, makes him understand that his life as a couple has suffered from some neglect and his marriage is going through a difficult time. All this happened while he, who was busy with work and daily chores, did not notice it.

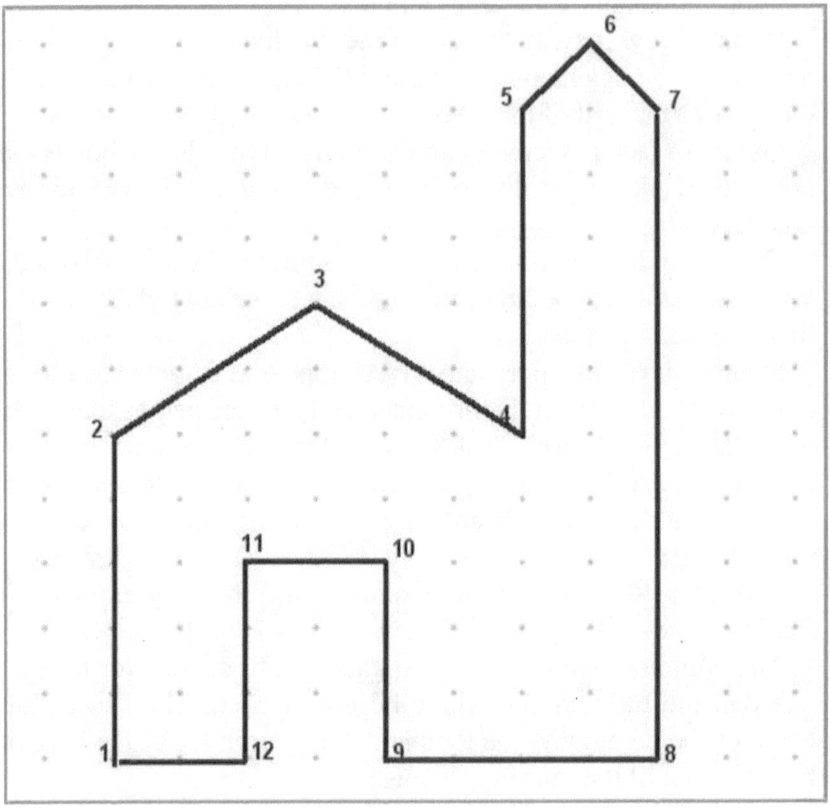

Figure 17 - Significant coincidences make it possible to materialize a "shape" in Philip's consciousness, as could happen in the game "Connect the dots".

If Filippo will be able to decipher the synchronicity of which he was the protagonist, he will probably be able to put his marriage back on better tracks.

Matthew and the frequency of the number 31.

The second example I present is based on the unusual frequency with which a number can come to our attention. We will talk about Matthew and his repeated encounters with the number 31.

One pale autumn morning Matteo, the sales manager for his company, was planning appointments with a number of clients. He had left the most important one last, the one with Dr. Corradini, purchasing manager of a company with a very interesting volume of business, but with an unfriendly and sometimes unpleasant character. Matteo was always uncomfortable in the various business contacts he had to have with him. With a sigh, he picked up the receiver and called him. Dr. Corradini was unusually affable that day, so that the interview soon came to its conclusion with the setting of an appointment.

- So doctor, can we meet on the 31st? - Matteo proposed.
- For so long? - Corradini objected.
- Well, it's the end of the month, a week to go.
- But look, you're wrong. We are in November, which takes 30 days. What calendar are you using? The Aztec one?

Matteo was confused for a moment, then he felt himself sinking. What a mistake, and then precisely with Corradini! He mumbled some excuses and set the appointment on 30.

The episode, however, left him troubled. How could he have so mistaken the date of an important appointment? It was obvious that November has 30 days. The well-known nursery rhyme also says it: “Thirty days has November, with April, June and September…“. And then, he had the agenda open on the table. How could he have come up with that cursed 31, exposing himself to the irony of the interlocutor?

For the rest of the day's work Matteo was not able to finish much, with that 31 that tormented him.

Finally the exit time arrived and Matteo took to the street to go to the bus stop, where he would wait for the 218 bus that would take him home. But here he found a sign: "Cause work in bus 218 is temporarily suspended. Alternatively, travelers can take the 31 bus that runs along Via Zanardelli ".

If he believed he got rid of the number 31, Matthew was wrong. He took this new vehicle which in any case led him near his home, but before going up to the house he went to the dry cleaners to take back some items of clothing.

Pina the mistress of the laundry, he co marked the clothes with the receipt.

- That's 33 euros.

Matteo wasn't stingy, but he liked to bargain whenever possible.

- So much? 33 euros? Come on Pina, I had a bad day, make me pay 30 euros.

- Are you joking, Mr. Matteo? I'd show you the electricity bill that arrived today! And then the government invents a new tax every day. Look, Mr. Matteo, I cannot make any discounts on the price. However, only for this time, I can only charge you 31 euros. More I can not take away, believe me!

Matteo paid and went out. Only when he was back on the street did he realize that the number 31 had once again entered his life.

Throughout the evening he kept mulling over that strange insistence of the number 31. When he got into bed he found he couldn't sleep.

The bad figure with Corradini burned in a particular way. Matteo was affected by the irony of his answer:

"But what calendar do you use? The Aztec one? ".

Yep, November has 30 days. Perhaps he had confused with the previous month, October, which brings 31.

- October 31st. What reminds me of this date? - Matte said to himself.

Of course, Halloween! (Fig. 18)

He began to think back to the Halloween nights spent with his friend Fulvio..

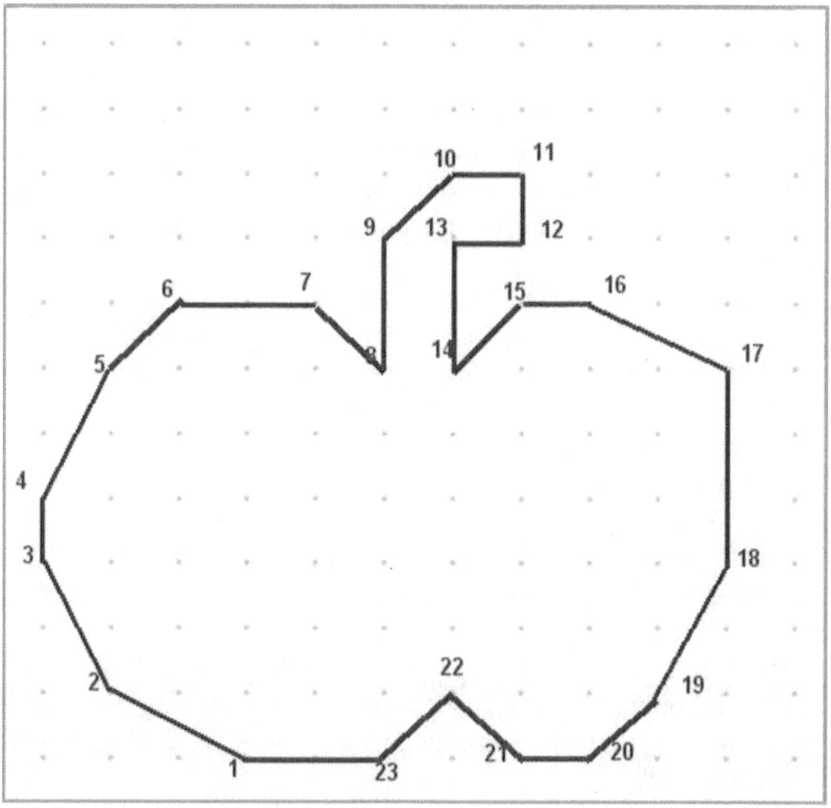

Figure 18-The game "Connect the dots" as a metaphor for the synchronicity that involves Matteo and, through the symbolism of the pumpkin, leads him to focus his attention on the Halloween parties of his youth.

Every year, on Halloween night, the two placed a large empty pumpkin with a lit candle inside, at the mouth of a bridge. Then they waited for passers-by, uttering screams and cackles to frighten them.
Fulvio was a few years older than him and was a true master in stealing the most suitable pumpkin in the surrounding countryside.
Matteo's mind wandered through these memories for a long time.
He hadn't heard from Fulvio for a long time. Who knows how he was doing it?
He looked at the watch. It wasn't too late to make a phone call.
From an old diary he retrieved Fulvio's number and called him.
Fulvio's receiver rang with incredible power in the silence of his house. He, sitting at his desk, was dumbfounded. The phone rang again, peremptory, overbearing.
Fulvio put the pistol he held in his right hand on the table and picked up the receiver..
- Hi old man, how are you? You know me? I'm Matteo.
He was dumbfounded. He pushed aside the paper and pen he had in front of him and replied:
- Of course I recognize you. Why did you think about calling me?
- Well, I thought we haven't seen each other for a long time and it would be nice to meet to exchange some ideas. What do you say, can it be done?
Fulvio stared at the receiver, as if amazed. He was no longer alone, someone had thought of him, the angel of the night had come to hold out his hand. Suddenly his loneliness seemed less overwhelming, and a whole film of disappointment and bitterness rewound in his mind bringing him back to the point where the crossroads between hope and despair was still viable, the choice could be pondered again. Yes, the foot on the edge of the ravine could still be pulled back. A new decision was possible.
Fulvio decided with the speed of lightning. He took the gun and put it back in the drawer.
"Of course it can be done, my friend ... yes, my friend, it can certainly be done, let's meet!".

The story of Matteo and Fulvio ends here. Nobody will ever know that Fulvio was saved that evening thanks to a series of strange coincidences that involved Matteo, all linked to the number 31.
No one will ever know which procedure (certainly not logical) led Matteo to transform some generic references to the number 31 into a significant date, the recurrence of Halloween, then arriving at the decision to contact his childhood friend.
Nobody will know how many other times stories like this have really happened.

The number, the archetype of the order.

"Not everything that can be counted counts
and not everything that matters can be counted. "
(Handwritten sign in Einstein's studio)

Numbers and synchronicity.

Numeric coincidences manage to fascinate everyone, especially when they are repeated regularly over a short period of time, even in different situations: the same time, the same house number, the same booking number, and so on. Many are struck by it, but not all believe there is a meaning or try to find out.

The repetition of the same number often conceals a very personal message, which must be intuited by digging into the folds of one's unconscious. The intrinsic symbolic meaning of the number itself, according to ancient traditions, can help in this research. Through a numerical allegory we can be protagonists of a synchronicity that tries to reveal a content of mysterious, perhaps divine origin.

The practice of establishing meaningful coincidences based on numbers is rooted in the most remote history of man. One could count the number of birds in the flock, the number of stones of the same shape or color, the number of plants in the same clearing. The number itself is particular and mysterious, that is, according to Jung, it has a meaning that is always numinous, strange, divine.

Some math teachers instructed their disciples in this way: if we take any group of objects and strip them of every single property such as weight, shape, size and so on, only one we will not be able to eliminate: their number. The number is an unavoidable quantity.

There is a close relationship between number and synchronicity, because both are clothed with sacredness and mystery. The number is always present in any synchronicity, in any form: quantity, date or other; it is often hidden under expressions such as "so much", "little", "very great", "infinite".

Each object can be considered in its unity or in its multiplicity. It is the number that establishes a precise order, therefore it creates hierarchy, order, coherence, organization, ultimately clarity.

The number is considered as one of the main archetypes, and Jung defines it as the archetype of the order. In fact, it creates discipline in chaos; it allows us to establish and grasp a regularity that exists a priori of our examination, even when, upon initial verification, it would be absent.

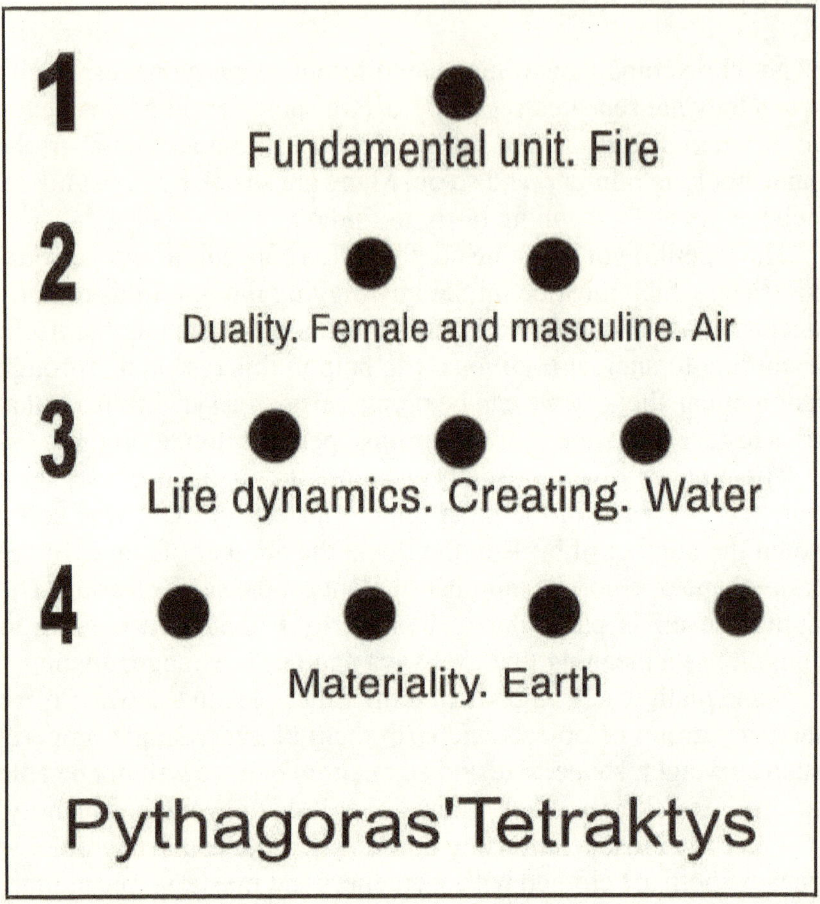

Figure 19 - The Tetraktys or quaternary number represented for the Pythagoreans the arithmetic sequence of the first four natural numbers, or more precisely positive integers.

The number is then the element that creates order in the spirit since the dawn of humanity, therefore the closest and the most intuitive for human intelligence. An example can be represented by the drawings of hands found in paleolithic caves. Along with paintings depicting bison, horses, rhinos and other animal species, handprints made with yellow or red ocher have also been found. The best known are those located in the "Cue-vas de las manos pintas" (Cave of the painted hands) located in Patagonia, in Rio Chubut.

Dating traces these handprints to at least 14,000 years ago. These are hands with well-spread fingers. Beyond the message that the artist wanted to communicate, and which, according to anthropologists, we will probably never know, the representation of the number five is very evident to the eye. As usually happens and as Occam's Razor suggests, the simplest and most direct interpretation is often the most correct.

In the most ancient tradition all the numbers have assumed a mystical meaning and a sacredness which is still preserved today; in particular, it is believed that all numbers from 1 to 9 are "sacred".

The Tetraktys of the Pythagoreans.

Pythagoras, Greek philosopher and mathematician who lived in the sixth century. BC, he intended mathematics as the science capable of putting order in the apparent chaos of the world. He founded a school in Crotone; to be admitted the students had to swear on the Tetractys: "I swear it, by the one who transmitted to our soul the Te-tractys in which the source and the root of the eternal nature are found...".

The Tetraktys represents the arithmetic sequence of the first four positive integers (Fig. 19). These numbers are arranged in the geometric form of an equilateral triangle with side four, thus forming a pyramid that summarized the relationship between themselves and ten. In fact, according to the so-called theosophical sum, $1 + 2 + 3 + 4 = 10$. In turn, 10 refers to unity by adding $1 + 0 = 1$.

The co-smogonic principles and the four elements of nature correspond to the four levels of the Tetraktys.

The golden section.

Some numbers are really special, like the number 1.618 for example. This value indicates the ideal proportion (note for example the proportion between width and height of a standard photo frame) and its formula is:

$\sqrt{5} + 1) / 2 = 1.618$ ...

It is normally indicated with the Greek letter Fi Φ and is called the Golden Number, or Golden Proportion, or Golden Section, or Constant of Phidias. Plato, who mentions it in his Timaeus work, speaks of it as the physical key of the Cosmos. In the Middle Ages the golden proportion was taken up by Leonardo Fibonacci from Pisa, for which today it is known especially as the Fibonacci series (fig. 20).
The Fibonacci series is a sequence of positive integers in which each number starting with the third is the sum of the previous two and the first two are by definition equal to 1.
Therefore, the series is represented by numbers:

1, 1, 2, 3, 5, 8, 13, 21, 34, 55, 89,144, 233, 377, 610 ...

and can go on indefinitely. Indeed, Fibonacci's intent while developing this series was to find a mathematical law that describes the growth of a rabbit population. He did not imagine he would discover one of the marvelous natural rules of the Cosmos. In fact, the Fibonacci series is present in practically all the elements of nature and reveals a project of incredible elegance.

The relationship between the length of the arm and the forearm of an adult man is "golden", as well as between the length of the leg and its lower part. The same golden ratio is also found between the phalanges of the middle and ring fingers.

The mathematical and geometric properties of the golden section that are repeated with incredible constancy in the most diverse natural contexts are a confirmation of the existence of a close relationship between a "Creative entity" and creation. Chance would certainly not have led to such a regularity of perfection and beauty.

When the number reveals a content such as the golden section, it proves to be the instrument and mediator of a relationship between the whole and the part, as well as between each larger part and the respective smaller part.

No one denies the existence, in this natural configuration, of an ideal of beauty and harmony, which amply justifies the names of "Aurea" and "Divina".

In botany, perhaps the most significant example of an effective presence of the golden section is represented by the geometric arrangement of the leaves and inflorescences of some plants.

In the mid-1800s Auguste Bravais, a French botanist, in a series of studies conducted with his brother Louis found that in many plants the leaves are arranged on the stem according to a spiral in which the angle between two successive leaves is almost constant and is of about 137.5°, corresponding to the golden angle. This arrangement guarantees optimal access and enjoyment of each leaf in sunlight.

But there are many examples present in nature. For example, the number of flower petals usually equals a Fibonacci series value. For example, very many, such as irises and lilies, have three. Dog rose has five or eight, like ranunculus and frangipane (also called plumeria). Many daisies have 13, so the game "loves me, loves me not" almost always ends with the consoling "loves me"; this also happens when instead of the daisy you use the more common "meadow daisy", which has 21. The composites, also called asteracee, a vast botanical group that includes among other things sunflowers, gerberas and chrysanthemums, have 34, 55 or 89, while the common chicory, which belongs to the same family, has 21.

These are all numbers that are part of the Fibonacci series. Amazing how natural expressions that one would like to have been generated by chance, all have mathematical features inspired by a single source.

In addition to human anatomical forms, as already mentioned, also those of many animals such as sea stars are inspired by the golden section; most obviously spiral-shaped creatures do so, as suggested by the shell of the Nautilus. But that's still not enough: at this point it should no longer surprise us that the shapes of the Milky Way and other galaxies are also inspired by the golden section.

In the field of medicine, important discoveries have been made in the study of human movement. For example, in 2013 it was possible to confirm that in walking the relationship between the "stance" phase (support) and the "swing" phase (when the foot advances without touching the ground yet) is equal to the golden ratio. This is supposed to have the purpose of making the step proportional to facilitate its control by the brain.

The curse of the ninth symphony.

In fact, many associate the number nine with death. This negative correlation began to take hold when Gustav Mahler died in 1911; he had just completed the ninth symphony "Das Lied von der Erde" (The Song of the Earth) and was quickly putting his hand to the tenth, hoping to have escaped the fate of Beethoven who died soon after completing his Ninth symphony (Choral Symphony).

The curse of the Ninth Symphony was born and fed between 1800 and 1900 and still persists with reference to the fact that many composers died after composing it. Among the most famous victims, besides Mahler and Beethoven, we can mention:

- Louis Spohr, or Ludwig Spohr (Braunschweig, April 5, 1784 - Kassel, October 22, 1859) German composer, violinist and conductor.
- Franz Schubert (Vienna, January 31, 1797 - Vienna, November 19, 1828) Austrian composer and pianist of the Romantic period.
- Antonín Dvořák (Nelahozeves, 8 September 1841 - Prague, 1 May 1904) Czech composer.
- Ralph Vaughan Williams (Down Ampney, 12 October 1872 - London, 26 August 1958) British composer, author of symphonies, chamber music, operas, choral music and soundtracks.

- Joseph Anton Bruckner (Ansfelden, 4 September 1824 - Vienna, 11 October 1896) Austrian composer.
- Kurt Atterberg (Göteborg, 12 December 1887 - Stockholm, 15 February 1974) Swedish composer, conductor and music critic.
- Elie Siegmeister (New York, January 15, 1909 - Manhasset, March 10, 1991) American composer, music teacher and conductor.
- Roger Sessions (Brooklyn, December 28, 1896 - Princeton, March 16, 1985) American composer, music critic and academic.
- Egon Joseph Wellesz (Vienna, 21 October 1885 - Oxford, 9 November 1974) is an Austrian musicologist, composer and teacher.
- Peter Mennin (Erie, Pennsylvania May 17, 1923 - New York June 17, 1983) important American composer of Italian origins.
- Sir Malcolm Henry Arnold (Northampton, 21 October 1921 - Norwich, 23 September 2006) British composer.

In an essay on Mahler, Arnold Schoenberg wrote:

> "It is as if the Ninth were a limit. Whoever wants to overcome it must die. It seems as if something is being communicated to us through the Tenth Symphony. it is something that we cannot know now, that we are not ready for. Those who wrote the Ninth Symphony have come too close to what we call the "Beyond"".

The Club of 27.

The English expression "Club 27" refers to artists, mostly rock singers, who have died at the age of 27. The expression "Curse of J27" is also often used to refer to the fact that their name or surname begins with the letter J.

This belief arose following the deaths of Jimi Hendrix, Janis Joplin and Jim Morrison. Many believe that the first artist to fall victim to the curse was Robert Johnson, who (it is said) sold his soul to the devil for success. Others award the Club's First Member Medal to Alexandre Levy, who died aged 27 in 1892.

In 2009, on the occasion of the 15th anniversary of Kurt Cobain's death, Robert Smith of National Public Radio declared: "The death of these rock stars at the age of 27 has truly revolutionized the way people look at rock."

Indeed, the deaths of so many artists at such a young age can be traced back to alcohol or drug abuse, car accidents or even suicide.

Here are some members of Club 27:

- Jim Morrison, (8 December 1943 - 3 July 1971), poet, author and cash, officially died of cardiac arrest (but an autopsy was never performed).
- Amy Winehouse, (September 14, 1983 - July 23, 2011) singer-songwriter, died of alleged alcohol abuse.
- Brian Jones, (February 28, 1942 - July 3, 1969) founder of the Rolling Stones. He drowned in the pool of his villa.
- Janis Joplin, (January 19, 1943 - October 4, 1970) lead voice and songwriter for the Big Brother and The Holding Company. She died of a heroin overdose.
- Jimi Hendrix, (November 27, 1942 - September 18, 1970) Pioneer and virtuoso of the electric guitar. He died of suffocation from vomiting following the ingestion of alcohol and tranquilizers.
- Kurt Cobain, (February 20, 1967 - April 5, 1994) singer and guitarist of Nirvana. Died by suicide.
- Kim Jong-yun, (8 April 1990 - 18 December 2017) composer and singer of the K-Pop group SHINee. Died by suicide.
- Robert Leroy Johnson (May 8, 1911 - August 16, 1938) American singer-songwriter and guitarist. Died of unspecified causes.
- Cecilia, Evangelina Sobredo Galanes (11 October 1948 - 2 August 1976), Spanish singer-songwriter, dedicated to the pop genre. She dies in a car accident.
- Chris Bell (12 January 1951 - 27 December 1978) singer and guitarist. He died in a car accident.

- Alexandre Levy (November 10, 1864 - January 17, 1892) Brazilian, pioneer of the fusion between classical composition and popular rhythms of Brazil. He is believed to be the first known member of Club 27. He died of unspecified causes.

Wolfgang Pauli and the number 137.

Wolfgang Pauli, physicist and Nobel laureate, had a long relationship with Carl Jung, initially for therapeutic purposes and then for scientific interest. There is documentation of a very long correspondence between the two scientists. Pauli would have liked to create a psychophysical theory that would unite matter and psyche. After all, Pauli continually grappled with his dreams that seemed to be an effective guide for him. Despite the scientific mentality, he was not exempt from extrasensory manifestations: in the university environment in which he worked the so-called "Pauli effect" was well known, according to which he could not enter the laboratories of his colleagues without some equipment failing, or some test tubes breaking up.
Throughout his scientific life he was almost obsessed with a complex problem in physics, the fine structure constant, equal to about 1/137.
Despite his efforts, Pauli never managed to find the reason for that value. In 1958 he was admitted to the Zurich hospital, as suffering from pancreatic cancer, which led to his death in a short time.
With great surprise he found himself hospitalized in room 137 and immediately pointed this out to his assistant Charles Enz, when he went to see him. He also told him that he would never leave that room again, just as he was unable to solve the problem of his life. He judged that the number 137 was like a trap for him, and it was indeed so.

Figure 20. Leonardo Fibonacci's golden section..

The number 108 in oriental culture.

The number 108 originates from 12x9 and appears constantly in oriental culture. In particular, here are some occurrences of the number 108 or of its multiples and submultiples:

- It is the number of beads of the Mala or "Indian rosary", and of the Buddhist Akṣamālā.
- It is the number of sins in Tibetan Buddhism.
- Hindu deities have 108 names which can be recited by counting the 108 beads of the Mala.
- The Rig Veda has 10,800 verses.
- The Tibetan sacred books of the Khagiur consist of 108 volumes.
- In Śrīmad Bhāgavatam, Krishna is described dancing with 108 shepherdesses and subsequently marrying 16,108 wives.
- In Shivaism, Shiva Nataraja is depicted performing his cosmic dance in 108 poses.
- In Japan, at the end of the year, a bell rang 108 times to greet the new year.
- The temptations that a person must resist to reach Nirvana are 108.
- The temple of Baalbek, in Phenicia, had 54 columns.
- In the holy city of Lashanmjh, in Tibet, there were 108 temples.
- Angkor Vat, Cambodia, has 54 towers, 108 statues on either side of the driveway and 540 statues of divinities.
- The stars considered sacred in Chinese astrology are 108.
- The length of the antediluvian reign in Sumerian mythology was 108,000 years.

At the level of curiosity, we add that 108 is the number of suitors of Penelope, wife of Ulysses, in Homer's Odyssey.

The number 666.

We cannot finish this review on sacred or significant numbers without mentioning the number 666. It appears in a passage from the New Testament, in the Apocalypse of John, and refers to a "Beast" that rises from the sea and devastates the Earth. In particular, in contemporary times, the number has become a symbol of the Devil.

As a curiosity, I report a news released by the Associated Press on November 2, 1996. Seeondo the agency, 436 residents of the city of Madison had agreed to play the combination of the three numbers 666 in a local lottery called Luck Devils, on the occasion of the Halloween party. Unbelievable but true, the players won the sum of 1800 dollars. From the spread of this news it seems that many others have tried.

The determinism of death and the resurrection.

"A mature and expert philosopher who asks him for advice on the books of philosophy to study, should answer: what philosophy do you want? Because I already know what you are looking for. You want a philosophical doctrine that shows you what you believe and you want to keep believing.
What do you want then? God or atoms? Freedom or Determinism? Tell me. I have here, in my pharmacy, the jars that contain each of these things.
I can give you one or the other, depending on your wishes or needs. "
(Giuseppe Rensi, philosopher and lawyer)

Divine justice in Drynor Township.

In 1836 Davy Crockett, a folk hero of the American Far West, while traveling to join the Texas Army in San Antonio, camped in a grove on the shore of Honey Grove Creek, a small river just 11 miles long that runs through the county. of Fannin in Texsas. That evening he wrote a letter in which he enthusiastically described that place, called Honey Grove for the abundance of combs that wild bees created in the hollow trees.

Crockett later promoted the birth, in that place, of a town with the same name. In 1842 his friend Samuel Erwin, the first settler, settled there and Crockett himself celebrated his marriage. Subsequently the locality became a small center in expansion until it was constituted a city in 1873; trade, cultivation of cotton and other activities allowed it to grow to the level of a few thousand inhabitants, which favored the existence of a local four-page newspaper, the Honey Grove Signal.

The March 31, 1905 edition of this newspaper reports the details of a fact that has become a classic in the literature of strange and mysterious events, generally known as the case of the predestined bullet. There are several versions, but the newspaper's chronicle allows us to reconstruct the most truthful one.

The article appears in a column on the second page, with the title "More surprising than a film. A bullet avenges a girl twenty years after her death ".

The opening words of the article ("Amarvelous case of punishment on earth for the sins of the flesh occurred ...") well describe the cultural environment in which the event took place, and the interpretation that was given of it.

The article presents the news of the accidental death of Henry Ziegland, a resident of Drynor Township, a hamlet surrounding Honey Grove. The chronicle of his death would be of no interest if we did not link it to an antecedent dating back to twenty years earlier..

In 1885 Henry Ziegland was a wealthy young man devoted to all the pleasant activities typical of young people of all times, such as the courtship of girls.

Unfortunately, for him, the environment and current morals were not inclined to favor passing adventures so that, when a local girl, Maysie Tinchor, was courted, he gave in to his flattery and promises because he had a project of marriage. Henry didn't think so, however, and once he had the fruit of her wishes he abruptly liquidated her.

Maysie's pain and the knowledge that she had been deceived were immense, so much so that the poor thing took her own life. Of course, she didn't end up like that. Filford, Maysie's young brother, upset over his sister's suicide, decided to confront Henry to express all her contempt.

He then rushed to his house but, with excessive impulsiveness, took with him the family weapon, a large military musket capable of exploding a single bulle

HONEY GROVE SIGNAL.

Honey Grove, Texas, Friday, March 31, 1905.

DR. PRICE'S CREAM BAKING POWDER

Greatest Aid to Cookery

Ayer's Cherry Pectoral

Coughs, Colds

CROOK-RECORD CO.

PARIS, TEXAS.

The Fashion Event

Spring Opening

Millinery, Tailored Suits, Wraps, Skirts, Shirtwaists, Gloves, Hose, Corsets, Underw'r

Wednesday, April 5

Special Bargain Prices

Figure 21 - Edition of March 31, 1905 of the Honey Grove Signal, which reports in a column on the second page the case of the death of Henry Ziegland which occurred in circumstances so strange as to recall a history dating back twenty years earlier.

When the two found themselves facing Henry, older and more robust than the young man, he was not at all conciliatory. He offended Filford and Maysie. Also, he mocked the musket Filford was holding in his hands.

Blinded by hatred and desperation Filford lost control of reason, raised the barrel of his musket and fired the only shot in the direction of the opponent: immediately he saw a large stream of blood splashing from his head as he collapsed to the ground.

At that point, convinced that he had killed him, Filford ran away; he realized that he had completely ruined his family, adding to his sister's suicide the transformation of him into a killer. He armed the musket with another bullet and, in turn, took his own life..

Things, however, had not gone as he believed: the bullet had only grazed Henry's temple, and had gone to lodge itself in the trunk of the tree behind him. Henry, with the luck that often helps the wicked, was saved by suffering a superficial wound that healed in no time.

We do not know if in the following years he came to repent of all the evil derived from his behavior, probably not. He made his business prosper and married a wealthy widow with whom he had a son. This grew and, in anticipation of his marriage, Henry decided to expand the house to accommodate and keep the new family alongside, building a new wing right in the direction of the tree hit by the bullet twenty years earlier. To get the necessary space, Henry decided to cut down the tree. Let's follow the story of the Honey Grove Signal:

> "Henry and his son cut down the tree into which Filford Tinchor's bullet had penetrated, and tried to split it into pieces. The wood was so hard that the axes and wedges had no effect, so they decided to insert small charges of dynamite into the natural holes in the trunk. At one point, an explosion threw open the trunk of the tree releasing the old bullet. The force of the dynamite pushed the bullet out.

> With extraordinary force, the bullet reached Henry Ziegland's left temple and caused his instant death."

The editor who started the article with the words "A wonderful case of punishment on earth for the sins of the flesh" concludes: "In the tomb of the Tinchor brothers there is now only dust and silence. However, the bullet, with which Filford wanted to avenge their wrongs, did his job today".

Does divine justice exist?

The Honey Grove Signal chronicle raises a question as old as the world. Is there a divine retributive justice? Should we believe in a "higher Entity" which intervenes according to criteria of fairness to rebalance the fate of the oppressed?

The question is highly debatable. In fact, even if this body were operating in this sense, we would first have to establish what its idea of justice was, and if this idea, if anything, should coincide with ours.

From the point of view of the philosophy of science this idea is completely absurd. Events involving humanity must be evaluated with the deterministic criterion.

Determinism identifies a physical explanation for any phenomenon, assuming that the cause-effect chain must always be valid. The consequence is that everything that happens is never due to chance, but is determined by the physical laws that dominate the universe. These laws are well known as fundamental interactions. It is the forces of nature that make it possible to describe physical phenomena, and to date basically four have been identified. They are the force of gravity, the electromagnetic force, the weak nuclear force and the strong nuclear force.

Divine justice cannot be framed in any of the four fundamental forces, therefore, for determinism, justice governed by a "higher Entity" that intervenes to reward good or bad actions is a completely absurd hypothesis. Henry Ziegland died simply because he had the unfortunate idea of using dynamite, it follows that the cause of it all lies in his imprudence and in his case. Point.

Moreover, on the basis of the deterministic principle, if divine justice really existed, it would have to intervene all the times, not just once in a while.

These considerations lead us to a crossroads: either divine justice does not exist, or the deterministic principle is false .

Randomness and the supernatural sign.

We can learn more about the concept by considering the physical healings that take place in miraculous places. Contrary to what many believe these places are not only Catholic.

For example, the Jamkaran mosque, sacred to Shiites, is located six kilometers east of the holy city of Qom, in Iran. The keepers of the mosque have established a commission since 1998 to examine the miraculous healings that occur in that shrine and which are listed on a special register.

Many "pagan" places should also be remembered, such as a large number of sources or rocks with thaumaturgical properties, such as many dolmens and menhirs to which propitiatory powers of fertility are still attributed today.

In the culture of the American Indians there are the Wheels of medicine that do not evoke so much the concept of healing the body but rather the spiritual healing and the power that is obtained from the knowledge of the secrets of the universe.

One of the oldest Medicine Wheel is found in the Big Horn Mountains in Wyoming, and is believed to have been made in an era ranging from 200 to 500 years ago. It is made with white stones arranged circularly on a diameter of 28 meters. The circular shape symbolizes movement, the passage of time, life and the seasons. At the center of the Wheel is a cross, a symbol of the Great Spirit and the cosmic Tree which has its roots on the Earth and branches stretched out to the sky and to the four directions.

There are also purely psychic places of healing, such as the fantastic territories destination of shamanic journeys, where the healer sets out to fight the forces that hold hostage the health of the victims affected by the most diverse infirmities.

Of course, there are also many Christian places, known for the miraculous healings that occur there. The best known is the French sanctuary of Lourdes dedicated to the Virgin Mary, to which I refer because a medical commission has been established there for a long time, also composed of non-believing scientists, which certifies inexplicable healings with absolute parsimony.

Many healings of serious or very serious pathologies take place through immersion in the water of the Grotto of Massabielle, which closely resembles the Christian baptismal rite with which, through a symbolic death and resurrection, one is washed from all sin and, by extension, from every disease since, as Saint Paul says, death entered the world because of sin.

I remember that even the Jungian archetype of water, while having a double value, in its positive aspects can be the bearer of life in arid and deserted regions of the Earth, and by extension in situations of physical and spiritual human suffering.

Lourdes can be defined as a "miraculous place" because the number of certified healings that take place there is statistically higher than those that occur in other places. Sandro De Franciscis, president of the Bureau des constatations médicales of Lourdes since 2009, said that in 150 years about 7,500 unexplained healings have been documented, even if only 70 of these have been recognized by the Catholic Church as miracles. The very significant statistical variation between the number of healings that took place in this place compared to that relating to different or "ordinary" places is evident.

The problem is, not everyone heals. Despite this, the inexplicable healings of Lourdes are sufficient to strongly undermine the materialist dogma of determinism: their extraordinary number suggests that there may be a flaw in that theory. Moreover, the objection advanced by many scientists, according to which miracles are not miracles, but normal consequences of physical laws not yet demonstrated, has no merit because in the same way believers can affirm that they are a direct consequence of the divine will, even if the existence of God has not yet been proven.

Of course, we must recognize that if all the people who immerse themselves in the Lourdes tubs were cured, then we would fall into a reciprocal form of determinism, that is, we would end up confirming the theory that we want to refute instead.

The explanation given for this contrast of concepts is quite simple: miraculous healings must be interpreted as signs, that is, as indicators of a reality that overcomes determinism but without denying it. If in his daily reality man is subject to the deterministic rules that make the world work, this does not mean that there is a higher level that can transcend these rules and that manifests itself occasionally, just the necessary to signal to the conscience of humanity its presence and to ensure that human presumption can immerse itself in a… bath of humility.

Therefore, the miraculous healings would not be due to yet unknown physical laws, nor to mere chance, but would be signs of the presence of a "Universal Consciousness" capable of transcending its own laws. This absurd concept may become more familiar to us when we examine the behavior of matter at the subatomic level, discovering that at that level of reality all the physical laws underlying determinism are quietly ignored. The same matter that at the macroscopic level, that is, at our level of existence, obeys the fundamental forces of deterministic physics, at its most intimate level ignores them, disavows them and distorts them.

It is therefore possible that determinism is simultaneously respected and violated. It is possible that deterministic laws assert with certainty that a certain incurable disease leads to death, and at the same time, at a different level, a nondeterministic force intervenes to disavow this principle and produce healing.

Since we live at the level of deterministic laws it is natural that these, at our level, are ultimately stronger.

There is a deterministic law that even miraculously healed people do not escape, because we know very well that, even if healed and restored, in the end they will still be doomed to death. As Prince Antonio De Curtis says in a famous composition by him:

Do you know what death is? ... it is a leveling tool.

> A king, a magistrate, a great man, passing the gate of the cemetery, lose name, power and wealth.

Francis of Assisi himself, singing his praise of death, affirms in a succinct tone:

> "Praised be my Lord for our bodily death, from which no human being can escape."

Almost all religions and many philosophies, however, contest the determinism of death by configuring transcendent realities, capable of placing life on a different and eternal level. The major Christian holiday is Easter, which celebrates the death and resurrection of Jesus Christ, a pledge of the future resurrection and eternal life to which all believers are destined.

Other religions believe in reincarnation, that is, in a cyclical resurrection that culminates with the attainment of perfection, a state in which death ceases to exist. According to the conception of classical Buddhism (600 BC) what counts is the liberation from the pain of existence. Happiness, that is, nirvana, is the end of continuous rebirths, nothingness.

In the texts of the rabbis, paradise is described as an enclosure of roses and myrtles, a spiritual place in the midst of which the Tree of life stands. In the Koran, pious Muslims of both sexes enter a garden of pleasures: the climate is good, water abounds and so are fruit, meat and honey.

Can we believe in the resurrection?

In the conception of many religions, it is possible to return to life after death and this idea is not as strange as it might seem. It has become fashionable, even among less believing scientists, to claim that "we are children of the stars". Indeed, there are no scientific doubts about the fact that our bodies are composed of atoms which, after having formed in the furnaces of the first stars, then spread throughout the universe, reconfiguring themselves in infinite expressions of creation as part of other stars, planets, or minerals and vegetables, and also as part of the bodies of living beings.

Each atom of our body has been part of countless other expressions of matter. As for the atoms, therefore, we could speak of "immortality" (at least as long as the universe lasts) or reincarnation. In a broader sense also of resurrection, if we consider the death of the organism of which they are part and the "rebirth" as part of a different organism.

Of course, this is not the type of resurrection that we would expect: we would all like to resurrect in a body that would maintain the characteristics and above all the awareness of the life lived previously. That is, we would like to transfer the consciousness of the previous body to the new body.

It depends on how we view consciousness.

According to the materialist conception, consciousness is only a byproduct of brain activity, an epiphenomenon produced by the electrochemical interactions that occur in the brain mass; consequently, when the brain dies and falls into decay, even the conscience, which would never have existed, "dies" with it.

According to many philosophical and religious interpretations, however, consciousness is pure psyche that transcends the brain. Not only is it not confined to the skull, but it extends into the cosmos interfacing freely and autonomously with other consciousnesses and with a universal Consciousness. In this sense, consciousness could also be the informational link that unites all the atoms of our body, or rather the project according to which it becomes possible that they arrange themselves and assume functions such as to obtain as a result our own body as it is, and not different, with two hands instead of four and with fingerprints different from any other human being.

If the atoms of our body retained the memory of their belonging to this project, this memory would be immortal, our consciousness would be immortal and in a certain sense our body would be immortal.

The problem of entropy.

After all, why do we die? It all depends on a law called entropy or, more technically, "second law of thermodynamics". Put simply, entropy means that if you break a glass you can then break each piece into further sub-pieces, and then each sub-piece into successive sub-sub-pieces, but you can never restore the glass to its original state of integrity. Similarly, if you prepare a scrambled egg, you will no longer be able to reconstitute the whole egg, especially when you have cooked it.

We all have a very strong perception of the directionality of time, and we accept that it only flows in one direction. We accept the "forward" direction of time, but struggle to imagine the reverse direction. This surely depends on the fact that the straight direction is the one in which causality operates. We take it for granted that causes precede their effects. An event cannot occur that has not been caused by another, which occurred previously. It will never happen that someone can pick apples without someone having planted the tree first.

Entropy is linked to the passage of time, and predicts that everything goes towards deterioration, since time proceeds in only one direction and it is not possible to "rewind the movie" to return to a previous situation; there is an implacable norm, precisely the second law of thermodynamics, according to which the entropy of a closed system always increases. Of course, the universe is a closed system.

This "return" can take place in several ways. In some cases, as in the resurrection of the flesh supported by Christianity, the deceased returns to life with the same spoils of the previous life. In other cases it is stated that the return to life takes place in the body of another living being, even of a different kind. Some religions predict the end of corporeality but the beginning of a new life in the form of pure spirit.

For the ancient Egyptians, real life was that after death, so the body was embalmed to protect it from corruption. This practice suggests that they believed in the resurrection to be carried out in the same body rather than in reincarnation in a different body. In fact, in the mythology of Ancient Egypt, the god Osiris killed by his brother Seth was resurrected by his wife Isis and became the king of the underworld and the judge of the dead.

Zoroastrianism, which was the most widespread religion in Central Asia between the 6th century BC. and the 10th A.D. based on the teachings of the prophet Zarathustra, it holds that the bodily resurrection of the dead takes place in view of God's final judgment on Good and Evil. After death, the soul of the deceased passes over a bridge while his actions, good and bad, are placed on the scales. When good deeds prevail, the bridge appears wide, otherwise it becomes as thin as a knife blade. The result decrees the destination of the soul in the Garotman or Paradise, or in the worst Place.

Greek mythology tells in several cases of characters who rise from the dead, sometimes acquiring immortality.

Tantalus tore up Pelops, his still child son, and set the table of the Gods to see if they would understand that it was human flesh. When Demeter had already eaten one shoulder, Zeus realized the deception, recomposed the limbs of Pelops and resurrected him, replacing the cutaway with an ivory one.

Hippolytus, son of Theseus and Hippolyta, died while driving his chariot but the doctor Asclepius resurrected him with the help of Artemis.

After Zeus had betrayed Hera with Semele, and after the birth of Dionysus, the fruit of her betrayal, Hera retaliated by killing her rival. However, once Dionysus grew up, he managed to become immortal, descended into Hades and resurrected his mother.

Reso, young king of Thrace ally of Priam in the Trojan war, was killed in his sleep by Diomede but his mother, the Muse Euterpe, obtained from Hades and Persephone to resurrect him and was satisfied.

One of the Platonic myths tells the story of a soldier from Pamphylia who fell in battle, named Er, who is resurrected by divine will while his body is about to be burned on the pyre. Hence, Er entered Hades even though he had not yet been buried, and was then brought back to life while retaining all memories of his previous life.

Platonism stated that souls pre-existed in a celestial world, and fell prisoner in a human body at its birth. To free itself, the soul had to submit to a purification process consisting of successive reincarnations. This Platonic thinking was influenced by older philosophical schools. Only after the third century BC the Neo-Platonists adopted a Hindu-inspired vision of reincarnation.

Mystery religions celebrated the cyclical rhythm of life, death and rebirth linked mostly to the agricultural seasons and the forces of Nature. These religions did not presuppose liberation from death but rather the continuous rebirth into a new life after each death.

The cynics, according to the tradition deriving from Diogenes of Sinope and with reference to Socrates and the demigod Heracles, practiced mystery cults that celebrated death and resurrection.

Manichaeism. Mani, a Persian philosopher who lived in the third century, founded a religion based on principles inspired by Christianity, Buddhism, Mazdeism and Gnosticism. This religion wanted to explain the origin, composition, and future of the universe. About the life after death he spoke of three different destinies for the Perfect, the Hearers and the Sinners. The souls of the Perfect, after death, would be received by Jesus, and would be purified by the Sun, the Moon, and the Star. Their "particles of light", released, would be transformed into divinities to ascend to the First Man and live next to him.

Catharism was a religious movement that developed between the 12th and 14th centuries especially in the south of France. Cathar theology was dualistic: the world of matter was seen as Evil, while Good could only be found in heaven near God. The life of the Cathars was therefore aimed at bringing the soul, conceived as a divine spark of Good, towards Heaven, freeing her from Evil.

The Cathars accepted the idea of reincarnation, consequently those who did not achieve liberation in one life would return in a new life to continue the battle towards Perfection. Reincarnation was not always necessary, but generally yes, considering that not everyone could be able to break the ties with Evil in a single life. This explains why the Cathars, persecuted by the Albigensian Crusades, spontaneously threw themselves into the pyres and allowed themselves to be burned singing.

Jehovah's Witnesses believe in a resurrection and a future reward for the righteous, which will be divided into two categories:

- the first consists of the small flock, made up of 144,000 chosen ones who will reign in Heaven together with Christ without a carnal body, but with a spiritual one;
- the second includes all the other servants of God who will live forever, with a new body, on an Earth transformed into Paradise.

The unjust, however, will rise again in this Paradise, but will always be subject to the second death.

According to the Islamic faith, humanity is doomed to death and the Day of Judgment. On this day Allah will order his Angels to sound the trumpets which, at the first blast, will cause the annihilation of every being. At the second ring all men, without exception, will resurrect in body, soul and spirit, to be judged and rewarded with Heaven or condemned to Hell. Judgment Day is also called Yawm al-qiyāma (Resurrection Day).

Reincarnation in Eastern Religions.

In some Eastern religions the resurrection often takes the form of reincarnation, since their dogmas hold that every living being, from the ant to the elephant, to cite a few examples, is the bearer of the spirit or soul of a deceased.

The classic form of the doctrine of reincarnation comes from India where it was elaborated and transcribed in Brahmanical writings, in the ninth century BC. about.

The concept was made clearer between the 7th and 5th centuries BC. from the Upanishads, a set of Indian religious and philosophical texts composed in the Sanskrit language. This allowed it to be adopted by other great Eastern religions such as Buddhism. After Buddhism, reincarnation was also adopted by Chinese Taoism, around the third century BC.

Reincarnation in Hinduism.

Hinduism is a religion widespread in eighty-four countries. It developed on the Indian subcontinent using the already existing

Vedism. Hindus called the cycle of rebirth Samsara. This is a Sanskrit term often depicted with the image of a wheel. In Hinduism and other religions such as Brahmanism, Buddhism and Jainism, it indicates the cycle of life, death and rebirth.

According to a later meaning, Samsara also indicates "the ocean of existence", that is, the material world permeated with pain and suffering, therefore earthly life. But earthly life is absolutely devoid of real substance. In fact, the world as we see and perceive it is only a mirage, an illusion, "maya". This illusion creates Avidya in man, a metaphysical ignorance, that is, a wrong view of earthly and otherworldly life.

Maya is an Indian religious and philosophical concept that definitively takes on the meaning of "cosmic illusion" in Vedanta thought. It embraces the whole phenomenal world, non-existent and illusory, as opposed to the only reality which is Atman.

When the soul leaves the now dead body, it rests for three eras and then migrates to a new body. Through this cycle his spiritual condition evolves. The form of the living being in which the soul is reincarnated depends on Karma, or on the good deeds performed. The more good deeds there are, the better the condition of the new life is. On the contrary, negative actions produce reincarnation in a lower form.

Hinduism is a path towards knowledge and salvation that leads to final liberation when the soul (Atman) separating from the body and the world, joins the absolute or soul of the world (Brahman). Samsara ends when the soul has reached Brahman and knowledge.

Vedanta is the Indian philosophical system of the Upaniṣhad, of which the most famous representative was Sankaracarya (circa 788-820). Vedanta systematically elaborates the doctrine of the One-All according to which the concept of plurality, inherent in the world that appears to our senses, is based on illusion.

Reincarnation in the Upanishads.

The Upanishads were the first writings in which the place of the "second death" was moved from heaven to earth. Ignorance of one's own individuality puts karma into action, the law of cause and effect of Eastern spirituality. Brihadaranyaka Upanishad states:

> "Now a man is like this or like this; according to how he acts and according to how he behaves, so it will be: a man of good deeds will become good, a man of evil deeds, bad. Man becomes pure with pure actions, bad with bad actions. A person consists of desires. And as is his desire, so is his will; and as his will is, so is his action; and whatever action he does, he will consequently reap. ". (Brihadaranyaka Upanishad 4, 5-6)

Reincarnation (Samsara) is the practical way by which good deeds are put to use. Therefore, the individual is obliged to enter a new material existence until all the karmic debt that he has accumulated has been paid.

Therefore, in the Bhagavad Gita the concept of reincarnation is referred to as a natural process of life to which all mortals are subject:

> "As the embodied soul passes in this body from infancy to youth and then to old age, so the soul passes into another body at the instant of death. The realized soul is not troubled by this change. " (Bhagavad Gita 2,13)

> "As a person puts on new clothes and leaves the used ones, so the soul takes on new material bodies, abandoning the old and useless ones." (Bhagavad Gita 2,22)

In the Puranas in chapter 5 all the different forms of punitive reincarnation are indicated, according to the sins committed.

Whoever has killed a priest is born again consumptive, whoever has killed a cow is reborn hunchbacked or demented, whoever killed a virgin is reborn as a leper, whoever has stolen food is reborn as a mouse, whoever has stolen grain is reborn as a locust.

The Upanishads are scriptures rich in poetry as it appears from these verses:

> "You are the woman, You are the man, You are the young man and the girl. You are the old man who moves painfully leaning on his cane. You are born with faces looking in every direction. You are the blue butterfly and the green red-eyed parrot. You are the cloud that generates the lightning, You are the seasons and the oceans. You never began, You are eternity and omnipresence. You are the One from whom all worlds are born.

Reincarnation in Buddhism.

Buddhism views reincarnation as a form of purification, not as a reward or punishment. The soul joins the "All Divine" through the merits accumulated in repeated reincarnations, but when this happens it loses its individuality and therefore its conscious ego.

According to Buddhism, no living being, and not even any god, escapes the law of death and subsequent reincarnations into beings directly inferior or superior to the state of their previous life. Death is a transformation of energy. After death, life does not disappear but turns into something deeper and more aware; in these conditions man continues to exist until a new reincarnation.

This fate is common to all living beings, not just to men. Even animals, and all forms of life, down to the simplest such as bacteria, are reincarnated. The concept of "becoming" proposes a reality in which nothing is created, nothing is destroyed, but everything is transformed.

Even for Buddhism there is a hell and a heaven. If man has done evil, after physical death he is subjected to long or medium-term penalties. Instead, good deeds in life allow abode in a divine world, which however does not last for eternity. Final liberation from suffering and passions is achieved only by attaining complete enlightenment or Nirvana according to the thought of the Buddha himself:

> "For countless lifetimes I have wandered
> looking in vain for the builder of this house.
> It is more painful than ever to continue to be reborn.
> Oh, builder! Now I have found you.
> You will never build this house again.
> All your boards are broken,
> The supporting beam is broken.
> My mind has achieved supreme freedom.
> All desire is extinguished. "
> (Buddha)

Therefore, for Buddhists the soul of every being is immortal, and it gives life to all the physical bodies into which he subsequently moves. Birth and death are moments in the universal renewal process and are part of a natural cycle.

Reincarnation in Taoism.

Reincarnation is a difficult concept to find in the aphorisms of the Tao-te Ching (6th century BC). Indeed, these do not contain overt references to reincarnation which, therefore, must have appeared later in Taoism. The Taoist doctrine states that "something" passes from one life to another but it is not specified who exactly this destiny is reserved for. In the Chuang Tzu, an important Taoist text of the 4th century BC. it is read:

Figure 22 - Tao in Chinese means "Way". In oriental philosophies the symbol of the Tao, also known as the "Yin and Yang", represents the path, the becoming of all things, which is realized with a movement that oscillates between two extremes.

"Birth is not a beginning; death is not an end. There is existence without limits; there is continuity without a starting point.

Life is the companion of death, death is the beginning of life. Who understands how they work? Man's life is a breath. If you inhale there is life, if you exhale there is death..

Life and death are companions to each other, so why should we be anxious?

The ten thousand things are one: some we consider beautiful because they are rare or otherworldly, others we consider ugly because they are dirty and rotten, but the bad and the rotten can turn into rare and otherworldly and the rare and otherworldly can turn into bad and rotten. Therefore, you must consider everything as one: the wise never ceases to evaluate the unity of all things. "(Chuang Tzu, 23))

The resurrection in Christianity.

In the Old Testament, the hope of overcoming death or resurrection is first spelled out in the books of Isaiah and Daniel. According to these prophets, for those who live in the present, and for the dead of past generations, death can be overcome thanks to the divine act of the new creation.

Your dead shall live; their bodies shall rise.
You who dwell in the dust, awake and sing for joy!
For your dew is a dew of light,
and the earth will give birth to the dead. " (Isaiah 26.19)

> "And many of those who sleep in the dust of the earth will wake up: one for eternal life, the other for contempt, for eternal infamy. And the wise will shine like the splendor of the expanse, and those who have led many to righteousness will shine like the stars, forever and ever. " (Daniel 12,2-3)

In the sacred books of Judaism some resurrections performed by prophets such as Elijah and Elisha are mentioned.

The resurrection in the New Testament.

The manifestations of a life of the body after death, in Christianity in general, can occur in three ways: the resurrection, the assumption into heaven and the theophany, or the sensitive manifestation of a divinity.

The prophecy of the resurrection is fulfilled with the descent of Jesus into the Underworld, in the three days following his death, to free the souls of those who had lived before his coming.

Faith in the resurrection is announced in the Gospel and reaffirmed in other biblical books, especially in the Acts of the Apostles and in the Letters of Paul, where it is stated that death is "the wages of sin."

For Christianity, the death of Jesus was not simply the sacrifice of an innocent, but the unique and fundamental event of salvation; soon Jesus would return and in his resurrection all believers would be resurrected. Already in apostolic times, however, while it was becoming aware that the return of Christ would not take place in immediate times, it became necessary to clarify the fate of souls immediately after their death, waiting for this return.

It was therefore imagined that the souls of those who had deserved Salvation went up to Heaven after a period of purification and in this regard the prayers of intercession for the dead were offered.

Figure 23 - The resurrection of Lazarus in an ancient Greek icon.

Christ, on his return at the end of time, would then pronounce the Last Judgment, followed by the resurrection of the "flesh" (that is, of the bodies, transfigured in the likeness of that of Jesus).

The New Testament offers many resurrection accounts. However, most people remember only one case besides that of Jesus himself, namely the resurrection of Lazarus. But the resurrections described in the New Testament are more than ten.

Unfortunately, we are never told what happens to the resurrected people. What do these people do after the resurrection and how do they continue their life on earth? Only in the case of Jesus' resurrection do the Gospels describe what happens. Jesus appears to the disciples who, however, do not recognize him by seeing his person, but by the gestures he performs.

Jesus shows them the wounds of the passion, but these do not seem to cause him any pain. In fact, Jesus invites Thomas, an unbelieving disciple, to put his hand in his side. Interesting detail: Jesus sits at the table and eats meals with them. In his Gospel, Luke describes the behavior of the risen Jesus

The prophecy of the resurrection is fulfilled with the descent of Jesus into the Underworld, in the three days following his death, to free the souls of those who had lived before his coming.

Faith in the resurrection is announced in the Gospel and reaffirmed in other biblical books, especially in the Acts of the Apostles and in the Letters of Paul, where it is stated that death is "the wages of sin."

For Christianity, the death of Jesus was not simply the sacrifice of an innocent, but the unique and fundamental event of salvation; soon Jesus would return and in his resurrection all believers would be resurrected. Already in apostolic times, however, while it was becoming aware that the return of Christ would not take place in immediate times, it became necessary to clarify the fate of souls immediately after their death, waiting for this return.

It was therefore imagined that the souls of those who had deserved Salvation went up to Heaven after a period of purification and in this regard the prayers of intercession for the dead were offered. Christ, on his return at the end of time, would then pronounce the

Last Judgment, followed by the resurrection of the "flesh" (that is, of the bodies, transfigured in the likeness of that of Jesus).

The New Testament offers many resurrection accounts. However, most people remember only one case besides that of Jesus himself, namely the resurrection of Lazarus. But the resurrections described in the New Testament are more than ten.

Unfortunately, we are never told what happens to the resurrected people. What do these people do after the resurrection and how do they continue their life on earth? Only in the case of Jesus' resurrection do the Gospels describe what happens. Jesus appears to the disciples who, however, do not recognize him by seeing his person, but by the gestures he performs.

Jesus shows them the wounds of the passion, but these do not seem to cause him any pain. In fact, Jesus invites Thomas, an unbelieving disciple, to put his hand in his side. Interesting detail: Jesus sits at the table and eats meals with them. In his Gospel, Luke describes the behavior of the risen Jesus:

> "And just as they were telling about it, Jesus himself was suddenly standing there among them. "Peace be with you," he said. But the whole group was startled and frightened, thinking they were seeing a ghost! "Why are you frightened?" he asked. "Why are your hearts filled with doubt? Look at my hands. Look at my feet. You can see that it's really me. Touch me and make sure that I am not a ghost, because ghosts don't have bodies, as you see that I do." As he spoke, he showed them his hands and his feet.
>
> Still they stood there in disbelief, filled with joy and wonder. Then he asked them, "Do you have anything here to eat?" They gave him a piece of broiled fish, and he ate it as they watched."

After all, in the last supper, while drinking from the chalice of wine, Jesus said:

> "Truly I tell you, he will never drink of the fruit of the vine again until the day I drink it new in the kingdom of God" (Mt 26:29)

suggesting, in a crudely literal and certainly questionable interpretation, that in his new life as a risen he would continue to enjoy the pleasures of conviviality. However, it is undoubtedly a "resurrection of the flesh".

The dualism. Entropy and synchronicity.

"If you look at the disaster of human history with a cold eye,
it is difficult to avoid the conclusion that man is afflicted
from an innate mental disorder that leads to self-destruction.
Murder within the species, on an individual or collective scale,
is an unknown phenomenon in the entire animal kingdom,
apart from humans and a few species of ants and mice. "
(Arthur Koestler, in The Observer magazine)

The examples of synchronicity examined above contained a decipherable message in terms of "friendly rescue". Filippo is guided to the awareness that his marriage needs a revision, while Matteo, unknowingly, saves Fulvio's life. It could be said that the two synchronicities intervened as positive rebalancers of negative situations, or rather that they acted as "referees" in a situation that envisaged two possibilities that could be identified as "good" and "evil" according to very widespread human criteria, or according to prevailing way of understanding what happens in the world.

In fact, the common thought of human beings, from a theoretical point of view and according to a largely demagogic optimism, always favors solutions that can avoid pain and suffering for others. This is especially true when "the next" is unknown. Instead, sometimes we do not scruple to wish all harm to many of those close to us. Therefore, good and evil can be present at the same time in the judgments of the same person, often in the same judgment.

From a generic point of view, the term "Good" indicates everything that appears desirable to individuals and such that it can be considered as the ultimate goal to be achieved in one's existence. The concept of "Evil" is typically opposite, but not separate, indeed, coexistent.

Therefore, good and evil are two sides of the same coin, absolutely inseparable; they are not only so in human judgments, but in all cosmological visions and almost in all philosophies and religions, with minor or major emphases. Good and Evil represent a duality.

The dualism.

The term dualism defines any religious, philosophical or even methodological and scientific doctrine, which provides for the existence of two opposite or irreconcilable principles.

In this book I have already referred to many dualistic principles, such as that typical of Francis Herbert Bradley's philosophy, relating to appearance and reality. Bradley is an English philosopher who lived from 1846 to 1924. According to his thought, the truth lies in the unity of the whole: there is no separation or multiplicity of things. The whole world of human experience is appearance; only absolute consciousness is real. Another type of duality much covered in this book is that relating to religion and science, a topic explored by William James, an American psychologist and philosopher of Irish origin who lived from 1842 to 1910.

In truth, the term dualism should be attributed to Thomas Hide, an English orientalist born in Billingsley in 1636 and lived until 1703.

In his work "Historia religionis veterum persarum" (History of the ancient Persian religion), referring to Zoroastrianism and Manichaeism, he defines dualism as the concept, present in both traditions, of constant struggle between Light and Darkness, that is, Good and Evil, experiences that they are present at the same time and are independent and opposite. Every aspect of human existence depends on the use of the Good of Evil.

In 1710 the philosopher Gottfried Wilhelm Leibniz published in Amsterdam his work "Theodicy" on the goodness of God, the freedom of man and the origin of Evil. In his work, Leibniz analyzed the dualism of Good-Evil in the light of the biblical book of Job, tracing the same origin of the universe back to the principles of Good and Evil. In Leibniz every negative event, such as cataclysms or the existence of unmotivated evil that strikes the innocent, has its explanation in the superior law of harmony which oversees the divine creation and which justifies every event in the light of a superior plan unknown to man. .

Even Pierre Bayle (1647-1706) in his "Critical Historical Dictionary", indicates the presence of Good and Evil as the material origin of the universe, albeit from a rationalist point of view.

In more recent times, Marxism recalls the superstructure-structure dualism, and in the psychological field Jung proposes the conscious-unconscious dualism. One of the most investigated dualisms currently is the mind-brain one, which I have already mentioned in this book.

A non-philosophical but purely physical form of dualism, which takes on a decidedly relevant aspect for the purposes of this book, is that born from the quantum physics experiments conducted in recent years. This is the wave-particle dualism, which is the basis of the incredible reality of quantum entanglement.

However, there is another form of dualism, which many scientists active in the field of quantum physics or philosophy hint at or shyly allow to transpire in their works, and it is the duality between the entropy of man, that is the human tendency to self destruction, and the synchronicity which, despite this, guides humanity towards progressively higher evolutionary stages.

The entropic man.

The entropy or second law of thermodynamics, one of the basic laws of physics, provides that under normal conditions systems left to themselves tend to become disordered and corrupt in direct relation with the passage of time. Every living and non-living thing wears out, decays and self-destructs. This is the sure end that, from the physical point of view, all beings will have to face; it is an inevitable process that has no return.

With less scientific punctuality we can imagine that, in addition to the physical entropy that concerns all matter, there is a psychic entropy that concerns humanity. In fact, every man has an innate tendency to self-destruction, and the tendency of individuals widespread in society can inexorably become the tendency of entire communities and peoples.

Man tends to self-destruction to a much greater extent than any other living being. The destruction of the ecosystem and the nuclear and bacteriological wars represent enormous dangers today, equally balanced by those of yesterday, such as the wars of religion and conquest, slavery and every other technique of murder and oppression, since the first time Cain he probably used a stone to kill his brother Abel.

The gratuitous cruelty supported by the arms race has always dominated human behavior, without realizing how the destruction of our neighbor is the destruction of ourselves: countless times we have risked extinction, and still risk it with greater certainty to because of the power of current weapons.

Many restrictive laws and religious commandments are designed to stem the self-harm of the human species: do not kill, do not steal, do not do to others what you would not want them to do to you. Despite this, man has within himself the germ of self-destruction, he seems to seek the death of himself and that of others. Despite being aware of this risk, man is unable to emend himself from this psychic entropy of his.

In his Letter to the Romans, St. Paul describes this situation well:

> "In fact, I know that good does not dwell in me, that is, in my flesh; there is in me the desire for good, but not the ability to implement it; in fact, I do not do the good that I want, but the evil that I do not want. Now, if I do what I don't want, it is no longer I who do it, but the sin that dwells in me. I therefore find this law in me: when I want to do good, evil is beside me. In fact, I consent in my heart to the law of God, but in my members I see another law, which wages war on the law of my mind and makes me a slave to the law of sin which is in my members. I am unfortunate! Who will free me from this body doomed to death? " (Rom 7, 18-25)

Entropy is not what it appears.

In fact, even entropy has aspects according to which it is not exactly what appears at first sight, that is, that "irreparable trouble" described by Boris Yellnikoff in the film "Whatever Works" with the phrase: "It's called entropy; it's like when toothpaste comes out of the tube and can't go back inside. "

We take it for granted that, at the level of our sensory experience, entropy functions exactly like a disintegrating force that acts in the direction of time. However, if we leave our level to enter levels of space and time that exceed our direct experience, we can give different faces and meanings to entropy.

Let's take an example with our planet, the Earth, by comparing what it was in its beginnings and what it is now.

The appearance of the primordial Earth, about 4.5 billion years ago, was absolutely different from the present one; its surface was continually torn apart by mighty volcanic eruptions and battered by persistent rains that would last millions of years, up to generating the oceans. Therefore, an almost infernal panorama of bare volcanic rocks illuminated by the incandescent lavas of the new eruptions. Certainly, a place where it would have been impossible to live.

Earth's very first atmosphere, composed mostly of hydrogen, was blown away by the solar wind, but the abundant and continuous eruptions formed a second atmosphere composed, among other things, of carbon dioxide, carbon monoxide, ammonia, water vapor and methane . This environment, with the help of lightning and ultraviolet radiation from the Sun, was an ideal cradle for the formation of organic molecules ready to evolve into increasingly complex organisms. So, life soon appeared; the oldest known traces date back to 3.8 - 4 billion years ago and are the ancestors of current bacteria, very simple and primitive single-celled organisms capable, however, of carrying out the fundamental activities attributed to living beings, that is to feed and reproduce.

But how was the development of these primordial beings possible starting from the simple organic molecules present in the atmosphere of four billion years ago? What was entropy doing while this was happening? Because, if we're to be honest, we have to recognize that it was a process of construction, not destruction.

It is true, the squalid and bare grayish volcanic rocks subjected to the law of entropy crumbled over the millennia, and entire mountain ranges disappeared and leveled off, so that any human observer could certify, in his experience, the destructive power of entropy. But if that observer had been able to free himself from the limited temporal constraints of his observation of him, extending it longer over the millennia, he would have seen the rock transform into smaller and smaller parts until it becomes earth; he would then see the earth spread out into vast plains and offer a favorable living environment. An observer would certainly have seen the waters erode and dig the rocks, opening their way and destroying the environment crossed, only to then transform into rivers capable of carrying fertilizing elements in the lands born from the disintegration of the rocks, allowing the development of herbs, plants, flowers. , insects and other living beings of all kinds, that is, of that nature that today we define as wonderful and which is the basis of our food cycle and our survival.

Let's face it: without entropy we would not be here today. Therefore, beyond what might appear to an occasional observation limited in time, the final action of entropy is not destructive but creative.

The history of evolution radically contradicts the principle of entropy. Evolutionists argue that the spontaneous combination of disordered, dispersed and inorganic atoms would have occurred in a limited time and in a precise order to form extremely complex molecules such as proteins and then DNA, up to the spontaneous creation of millions of living species with still structures. more complex.

The law of entropy as it is declined denies that this process can take place; but, in fact, it did. Evolutionary scientists question this problem without being able to propose solutions. Evolutionary scholar Roger Lewin states in an article in the journal Science:

> "According to the second law of thermodynamics, systems should decay over time, presenting less, not greater order."

Evolutionary scientist and physician Jesse Hackley Rush writes:

> "In the complex course of its evolution, life reveals a remarkable contrast to the trend expressed in the second law of thermodynamics. While the latter speaks of an irreversible progress towards increasing entropy and disorder, life continually evolves towards higher levels of order. "

The volume "The deception of evolution" (Harun Yahya) reports the opinion of another evolutionist scientist, George Stravropoulos, published in the well-known scientific journal American Scientist:

> "Under ordinary conditions, no complex organic molecule could form spontaneously, but rather disintegrate, according to the second law. In reality, the greater the complexity, the greater the instability and the greater the certainty, sooner or later, of its disintegration. Photosynthesis and all life processes, and life itself, despite the confused language, cannot yet be understood in terms of thermodynamics or any other exact science."

An "ab absurdum" solution is proposed by the famous evolutionist Jeremy Rifkin in his essay "Entropy", published in 1980, in which he speaks of a "magical power" that would allow evolution not to be overwhelmed by entropy.

In fact, a "magical power" capable of domesticating entropy exists, and it is synchronicity. As in the Evil-Good dualism the two forces coexist and create a cosmic reality destined, if it ever ends, for the victory of the Good, so the Entropy-Synchronicity dualism operates in the cosmos in a challenge that will see the winner, in the end, the Entity from which synchronicity emanates.

Synchronicity as an intelligent evolutionary project.

Synchronicity is an evolutionary project guided by an intelligence, which we can define as universal Mind, capable of intervening not only in the life of individual people, as in the cases of Philip and Matthew, but also in the life of groups, peoples, nations and humanity. whole. This justifies the fact that neither entropy nor human selfishness have prevented the physical universe from evolving towards ever greater perfection rather than disintegration, and the history of humanity from finding its end not in collective self-destruction but in a development of art, culture and, ultimately, the awareness of a destiny that goes beyond physical death.

In my previous book Entanglement and Synchronicity I dwelt on the truly disconcerting circumstance that the human species, which began millions of years ago with the first hominids, remained in the state of stone civilization for an immense time until, only twelve thousand years ago, it kicked off a lightning-fast cultural leap that has led us today to travel the solar system.

All of this is not possible in a gradual evolutionary process. Something happened twelve thousand years ago, something that, after millions of years of cultural slumber, unlocked human consciousness allowing today's peaks to be reached.

When it happens that some discoveries are made at the same time, in different places in the world, by two scientists unaware of each other's work, we must imagine that the input was given by a collective psyche, common to both. For example, this happened with the invention of the telephone by Antonio Meucci in New York and Alexander Bell in Boston.

On April 22, 1984, the Centers for Disease Control and Prevention publicly declared that the French Lav virus has been definitively identified as the cause of AIDS by researchers from the Pasteur Institute. The next day Margaret Heckler, the secretary of the Health and Human Services ", announces that Robert Gallo, director of the cell biology laboratory of tumors at the National Cancer Institute, has in turn isolated the virus candidate to be responsible for the disease from AIDS patients, calling it Htlv-III. Thus begins a a real legal battle between the two prestigious research institutes, both of which claim the paternity of the discovery.

It is beyond question that many discoveries or technologies such as arches, agriculture, animal husbandry or pyramid-shaped constructions were made by different and very distant peoples, practically "simultaneously" if we consider that they occurred in the last twelve thousand years on a history that has lasted a total of millions.

Today the main concern of individual scientists and research laboratories is to publish the results before someone else does; in fact everyone knows, even if they do not have precise knowledge of it, that surely someone else, somewhere in the world, is doing the same research as them. Incidentally, this produces real tragedies in the consciences of researchers, torn between the urgency of publication and the fear of not having considered some aspect that could affect the results.

Furthermore, we cannot fail to notice the numerous discoveries that occurred "by chance", as if a mysterious prompter wanted to point them out from time to time. We are talking about X-rays, super-glues, anesthesia, penicillin, the microwave oven, saccharin, transistor, velcro, Teflon, verdigris, and many others, such as Coca-Cola. This was invented on May 8, 1886 in Atlanta, (Georgia, USA) by the pharmacist John Stith Pemberton who absolutely did not want to invent a drink, but a remedy for headaches and tiredness, therefore a medicine. The first name that was given to the drink was "Pemberton's French Wine Coca".

Cultural synchronicity.

Joseph Cambray in his book "Synchronicity. Nature and Psyche in an Interconnected Universe" talks about cultural synchronicity by citing events that span entire civilizations and millions of people, such as the emergence of democracy in classical Greece, an event that developed over time very short and in a very small portion of the planet..

I believe that at the moment we are experiencing another great cultural synchronicity driven by the development of quantum physics, a branch of science born a few decades ago. Quantum physics is unveiling incredible possibilities, such as the existence of non-physical levels that interconnect all matter in the universe into a single network.

These are levels in which particles, massless and in the form of vibrating energy and pure information, can free themselves from the constraints of time and space to which matter must undergo in our level of understanding.

The "quantum synchronicity" we are experiencing should lead to an unthinkable cultural leap: the overcoming of materialism and the birth of a unifying science of nature and psyche, spirit and matter. Only through a psychic level will it be possible to overcome physical barriers and enter the Consciousness of the universe, to benefit from all the energy and intelligence contained therein. It will be a leap whose potential we cannot even imagine and which will open the doors of "heaven" to humanity with all that it contains.

The anthropic principle.

"We are the local incarnation of a Cosmos that has grown to self-awareness.
We have begun to understand our origin: we are stellar matter meditating on the stars "
(Carl Sagan)

"O resplendent Sun, what would you be, if I were not down here, to make you shine upon me?"
(Friedrich Nietzsche, Thus Spoke Zarathustra)

The role of man in the universe.

The idea that emerges from the reading of the previous chapter, according to which a "Universal Mind" would operate through episodes of individual or collective synchronicity for the cultural development of mankind, presupposes that man living on planet Earth is assigned a central role . To fulfill this role, man would be the object of attention, protection and guidance from this "Mind".

Conversely, in scientific cosmology, man represents nothing more than a product of the evolution of matter, absolutely irrelevant for the development of planet Earth, much less for that of the Universe. In truth, until the 16th century the cultural vision placed man at the center of the universe. The theory subsequently developed by Copernicus, according to which the planets revolve around the Sun, caused the Earth to become, from the center of the universe, a simple small planet. Consequently, man too became just one of the many living creatures in a tiny part of the universe.

The anthropic principle.

Nel 1973 uno scienziato, Brandon Carter, elaborò una nuova In 1973 a scientist, Brandon Carter, developed a new formulation of the relationship that unites the human being to the universe. He called it "Anthropic Principle" and presented it to the scientific community during a Symposium on "Comparison of Cosmological Theories with Observational Data" which was held in Krakow as part of the celebrations for the 500th anniversary of the birth of Nicolaus Copernicus. This principle, much contested in its so-called "strong" formulation, overturns the Copernican vision of the relationship between human beings and the cosmos and restores man's centrality.

Even before Carter made the theory public, it was already being shared with other scientists such as Martin Rees, Freeman Dyson, John Wheeler, and others.

Figure 24 - Commemorative stamps issued by the Polonian Post Office on the occasion of the celebrations for the 500th anniversary of the birth of Nicolaus Copernicus, during which Brandon Carter coined the term Anthropic Principle.

Figure 25 - Memory of Stephen Hawking, recently deceased, and his book "The universe in a nutshell". In the book, he deals with issues related to the M-Theory on the multidimensional universe and frequently refers to Brandon Carter's "Anthropic Principle".

It must be said that a scientist of the caliber of Stephen Hawking, at first doubtful, after the formulation of Theory M (theory that unifies everything in a multidimensional universe) changed his opinion by becoming a supporter of the Anthropic Principle and including it in all his subsequent studies and publications , such as the book The Universe in a Nutshell.

Brandon Carter.

Brandon Carter (fig. 26) is an Australian physicist, known for his studies on the properties of black holes and for the discovery of a fourth constant of motion, baptized, with reference to his name, "Carter's constant". He is therefore a scientist of caliber, and this made him particularly reliable when`he formulated the theory he defined as! Anthropic Principle ".

In its general lines, this principle holds that human consciousness is not the casual result of the evolution of matter, but the point of arrival of an evolution of the universe that tended precisely to the birth of consciousness. The Principle as Carter enunciated it in Krakow has two formulations, called "weak" and "strong".

The "weak" anthropic principle.

The so-called weak anthropic principle argues that our position in the space and time of the universe is certainly privileged because all the constants underlying the functioning of the universe are calibrated on the most suitable values for the development of our type of life. Due to a series of incredible coincidences, the laws governing the universe are consistent with the possibility that intelligent life exists. For example, if the Sun were of a slightly different size, or if the Earth revolved around the Sun at a minimally different distance, the seas would become boiling pots or mountains of ice. The Earth rotates with a very small eccentricity, equal to 2%, and the inclination of the rotation axis is such as to accurately allow the alternation of the seasons.

Life on Earth is based on carbon, which was poured onto our planet following the explosion of the stars in which it was formed.

igure 26 - Above John David Barrow and Frank Tipler, authors of the book "The anthropic principle". Bottom left Brandon Carter, who formulated the theory..

The four fundamental forces that govern the cosmos depend on some physical constants such as, for example, the Hubble constant, the Planck constant, the universal gravitational constant, the electric charge and the mass of the electron.

The fact that all these constants are exactly calibrated on values capable of allowing the development of life is incredibly surprising: it would be enough for one to have a minimally different value to make life impossible.

The possible values of the four fundamental interactions (gravitational, electromagnetic, weak nuclear and strong nuclear) and of the subatomic particles (proton, neutron, electron) would be infinite, and only the values we actually measure allow us to create an interactive network between them that gives rise to a universe that generates life. Among the infinite configurations. only one was chosen, very specific. Stephen Hawking writes in his book "From black holes to the universe":

> "If the density of the Universe one second after the Big Bang had been greater than the critical density by one part in a hundred billion, the Universe would have collapsed after 10 years. If, on the other hand, it had been less than the same value, the Universe would already be empty after 10 years of existence "

The "strong" anthropic principle.

The strong principle states that the universe must be just as it is to allow the development of life, at a certain moment in its history, which coincides with the epoch we are living in, that is, about 14 billion years from its formation. Among the many universes that could have been born from the Big Bang, the very one that allows us to be here today has developed; the probability of this happening is one in several billion; if it had happened by chance, then chance would have won an impossible lottery.

This means that the universe was designed precisely to allow us to be here today. The universe has one purpose: to host man.

Therefore, man would not be a guest who happened to be by a bizarre chance in a universe that ignores him, but the reason and purpose for which the universe exists. This theory implies that the universe was formed with the aim of promoting life and the development of consciousness; that is, that the universe was designed as a cocoon, a placenta in which life in general and human consciousness in particular could originate.

The assumption that the universe was created for us certainly assigns a purpose to the cosmos, which is no longer a chaotic mass of brute matter but responds to a project for the realization of man in a becoming no longer entrusted to chance, and it requires the presence of an "Intelligent Design", and therefore the existence of a "Designer" or in any case of that "Universal Mind" mentioned several times in the previous pages.

This could also have religious implications on which, however, there are no pronouncements, at least on the Catholic side. Father Gabriel Funes, Jesuit, former director of the Vatican Observatory, that is of the Astronomical Observatory of the Catholic Church, stated in this regard that "The Church has no theory and does not have a position on the anthropic principle, but remains open to what science can discover ". There are no known pronouncements by the current director, Brother Guy Joseph Consolmagno.

John David Barrow and Frank Tipler.

Physicists J. Barrow and F. Triple published in 1986, in the United States, the volume "The Anthropic Cosmological Principle" in which they not only deal extensively with the theory, but propose a significant study of it.

The two authors extensively describe the conceptual basis of the Principle: if there were not an alarming number of extraordinary coincidences in the form of physical laws and in the values of the constants of nature, biochemistry, life and intelligent life would not be possible. A universe created by chance would not allow life, and not even ordinary astronomical objects and ordinary matter would be possible.

Starting from this observation, they come to the conclusion that a project necessarily exists. In their study the weak anthropic principle, which was limited to the recognition of facts, evolves into the strong principle. Barrow writes in the book:

> "Trying to imagine a whole collection of hypothetical" other universes "in which all the quantities that define the structure of our universe take on all the possible permutations of values, we would discover that almost all of these possible universes we created on paper are stillborn, unable to generate the kind of chemical complexity we called "life". This discovery led Brandon Carter to suggest that there may exist, in the universe, some more speculative metaphysical aspect, which he called the "strong anthropic principle", to distinguish it from the more controversial weak anthropic principle. The strong principle states that since there appear to be such a large number of remarkable and seemingly disconnected "coincidences" conspiring to allow life to be possible in the universe, this must give rise to observers at some stage in its history. . "."

The ultimate anthropic principle.

Barrow and Tipler suppose a third form of the Anthropic Principle, the "Ultimate Anthropic Principle", according to which "An intelligent processing of information in the universe must necessarily develop and once it appears, this will never die out".

According to the two scholars, a universe with the ability to produce intelligent life would make no sense if this universe did not last long enough to be able to develop an intelligent species.

The "Participatory Anthropic Principle".

This form of the anthropic principle is suggested by John Archibald Wheeler as an alternative to the strong principle. According to Wheeler, observers are necessary for the life of the universe, as they are necessary for its knowledge; therefore man, as an observer of the universe, actively participates in his own existence

"Missionaries of Conscience"

In 2005 an article appeared on the SETI website (Search for Extra-Terrestrial Intelligence, a program dedicated to the search for extraterrestrial intelligent life) entitled "The Drake Equation" in which the participatory principle was further extended. Referring to any messages between inhabitants of the cosmos, the article stated that "The mission of the senders is to spread Consciousness in the Universe". In other words, whoever sends messages to the universe becomes a "missionary of Consciousness". He spreads "Consciousness", in the form he knows, to entities capable of benefiting from it.

The anthropic principle and time.

The most objective science does not reject the reasonableness of the anthropic purpose, as it would be decidedly absurd for such a perfect cosmic construction to derive from chance and have no purpose. This being the case, a frequent objection is also answered: "If the goal is man, who lives only on planet Earth, why does the universe need so much space and so much matter?".

The answer is that the constants that allow for life have taken time and space to materialize. It was necessary for the universe to age and expand to allow the stars to manufacture the chemical elements that are the basis of life on Earth, such as carbon. Moreover, the existence of other planets on which the same life has developed is not excluded, as well as the possibility that man can export his civilization to other planets and galaxies.

The first element that appeared in the universe was hydrogen, whose atom contains only one proton. About three minutes after the Big Bang, the fusion of two or three hydrogen atoms allowed the formation of helium and lithium, which have two and three protons respectively.

Inside the stars, the fusion of two helium atoms produced beryllium. From the fusion of a beryllium atom and a helium atom, unstable carbon atoms were born, which became stable only when the temperature in the star rose to 100 million degrees Kelvin. Finally, when the star, at the end of its life, exploded generating a supernova, carbon and other metals were ejected into space and went to form other celestial bodies, including planets.

At this point, having established that life is essential for the existence of the universe, and that the universe is probably infinite in space and time, we can ask ourselves if life also has the goal of becoming infinite in time and in space. And above all if this infinity (immortality) can be referred to life in general understood as the conservation of the species, or not even to every single individual individually.

Would it be possible to imagine immortal living beings?

We have already attributed immortality to the essential components of matter, the atoms. In the biological field, among living beings we can refer to bacteria, which usually do not end their existence with death but with splitting. Bacteria have the ability to divide into two parts, each of which retains all the genetic makeup of the original bacterium. Barring accidents, this process lasts indefinitely and it is easy to calculate that, if a bacterium breaks down every half hour, in a day it is capable of generating 250,000 billion similar bacteria.

If even higher organisms, such as man, reached the ability to split while maintaining the neuronal heritage, that is, self-awareness, they would practically become immortal. This, at the moment, is to be considered impossible: in fact, bacteria are made up of a single cell while the human body includes myriads of cells specialized in a different way for each organ (skin, liver, etc.) which should all split at the same time.

The Omega Point by Frank Tipler,

Frank Jennings Tipler (fig. 26) co-author with John Barrow of "The Anthropic Principle", is an American lecturer and writer who teaches mathematical physics at Tulane University in New Orleans. Most of his studies have focused on the physics of time travel.

He is also known for being the author of the "Omega Point" theory, which actually takes up the homonymous theory previously developed by the French Jesuit scientist Pierre Teilhard de Chardin.

Tipler tries to study the ultimate fate of the universe in a similar way to what Isaac Asimov does in his novel "Last Question". In his proposition, which he bases on reasonable hypotheses suitable to support it, he starts from the concept of "eternal progress" to hypothesize the future existence of a God, that is, a man or alien who has become immortal, resembling Nietzsche's superman. He also hypothesizes the resurrection of the dead by computer and "mind uploading". This technique would allow the recovery and transfer of an individual's mental patrimony from the old to a new body, or to the body risen from the dead.

Furthermore Tipler hypothesizes that, thanks to a supercomputer, the universe could be bent to the wishes of humanity, even acting on the time already elapsed; in this way it would be possible to reconstruct events, worlds and people of the past, with a technique of eternal guided return.

In fact, Tipler proposes a perspective linked to transhumanism, a cultural movement that advocates the use of science and technology to increase physical and mental abilities and to improve all undesirable or harmful aspects of the human condition, such as disease and aging up to a possible transformation that would go beyond the human condition.

Tipler's theses, as expected, have aroused much controversy in the scientific world. In a review of Tipler's book "The physics of immortality", South African physicist George Ellis calls the Omega Point theory "a masterpiece of pseudoscience".

Other scientists, such as physicist David Deutsch of the University of Oxford, consider Tipler's arguments to be consistent. In his book "The Fabric of Reality" Deutsch, exposing his theory of Everything, he incorporates Tipler's Omega Point.

At the spiritual center of the universe.

In summary, we can say that we are here because of the universe and the universe is here because of us. The Anthropic Principle is reconfiguring modern scientific thinking. This began with Copernicus and his discovery that the Earth and planets revolve around the Sun; with this, he had subverted the Aristotelian thought that placed the Earth and man at the geographical center of the universe.

Developments in modern physics are making a new reversal. This time man, although no longer the geographical center as Copernicus rightly demonstrated, firmly asserts himself as a spiritual or psychic center.

Second part. The confirmations.

But what is matter?

"Modern physics, therefore, represents matter not as passive and inert, but in a continuous dance and state of vibration, whose rhythmic figures are determined by molecular, atomic and nuclear structures. This is also how Eastern mystics view the material world. They all emphasize that the universe must be grasped in its dynamism as it moves, vibrates and dances; that nature is not in static but dynamic equilibrium. "
(Fritjof Capra, physicist and essayist)

Looking for evidence.

In the first part of the book I presented many theories elaborated and proposed by scientists, based on thought, philosophy and logical deduction but not supported by any experimental evidence recognized by the accredited community. An important first step to give scientific dignity to all the hypotheses discussed would consist in establishing an ascertainable link between the physical level of functioning of matter and a "spiritual", that is, non-deterministic level.

At the level of our experience, which is absolutely deterministic, it is difficult to find non-material connections: a table or a chair are pieces of wood or metal and do not inspire, as far as our senses appear, psychic or metaphysical implications.

This is no longer the case, however, when we enter the extremely small level, observing the behavior of elementary particles. When finally, in recent decades, man was able to penetrate subatomic reality, nature reserved for him incredible surprises from which it is possible to deduce the existence of an exclusively psychic level, made not of matter but of energy and intelligent information.

This second part of the book will be devoted to exploring the wonderful world of subatomic particles.

Albert Einstein introduced the concept of relativity into physics. Many certainties, deriving from our way of understanding the world, have collapsed when we learn that time and space are relative concepts and can have different values depending on the conditions. The "twin paradox" is paradigmatic: one of the two gets on a spaceship and travels through space for many years, at a speed close to that of light. When he returns home, he discovers that time has passed much slower for him than for the people left on Earth and many of these, who were the same age, are already dead: his own twin is many years older. .

Faced with this scientific evidence, we should not be surprised to find that matter is also relative. At the macroscopic level or medium-sized area in which we live, it generally appears well defined: its boundaries can be determined, it can be weighed and measured, its degree of hardness or malleability can be established. If not, we would be in trouble every day. These properties of matter remain unchanged even at an astronomical level, and the physical laws concerning the table in our kitchen also apply to the last planet of a distant galaxy.

At a subatomic level, however, that is, in the extremely small, things change and matter becomes something very unclear with respect to its true essence. The subatomic level is that of elementary particles, which are the internal components of the atom..

Philosophy and mysticism of physics.

The philosophy of physics studies the logical and metaphysical aspects of concepts such as energy, time, space and matter. She is particularly interested in issues of causality and determinism.

Of course, philosophy has been interested in quantum physics ever since it was born, due to its ability to undermine fundamental aspects of the scientific tradition, such as the concept of measurement, which was abruptly replaced by that of uncertainty.

The operation of the measurement at the subatomic level becomes impossible, as it is performed on elements so small that they are perturbed by the measurement itself. Consequently, atoms cannot be observed directly even with the most powerful microscope, but only "theorized".

With this the failure of determinism so dear to Pierre Simon Laplace is sanctioned and the birth of new values such as probabilism and indeterminism is announced.

Indeterminism is a philosophical attitude, headed by the English astronomer Arthur Eddington, who admits the action of chance in the outlining of will or reality, that is, he admits that things can also happen by chance, for no apparent reason, and not only as a product of cause and effect. The notion of indeterminism has a clear definition:

> "The present state of the physical system is not completely definable, or the same completely defined present state can correspond to many possible future states, only one of which will be realized."

Among the supporters of indeterminism are the greatest physicists of our times: Werner Karl Heisenberg, Niels Bohr, Arthur Compton, Pascual Jordan, Hans Reichenbach, Robert Millikan, Georges Lemaître and others.

Quantum physics comes out of deterministic rigor and leads us to a substantially mystical conception of the cosmos, based on elements that recall a possible spiritual component alongside the material one. Thinking about the reality of things goes back to over two millennia ago, to the intuitions of the first Greek philosophers. However, contrary to what happened at that time, the vision of the cosmos that is emerging today is not based on intuition but on experiments conducted with great precision in the best physics laboratories.

The roots of all Western science and physics in particular are to be found in the sixth century BC, in a period that knew no division between religion, philosophy and science, among the wise men of the school of Miletus. Their studies were aimed at discovering the real constitution of matter, which they identified in the term physiká, hence today physics. The physiká represented the effort aimed at discovering the essence of things. On the basis of their assumptions, the philosophers of Miletus were defined ilozoists, a term that means "those who think that matter is animated". In fact, they made no distinction between spirit and matter; indeed, they did not even have a term that defined matter, as for them all forms of existence were endowed with life and spirituality. In particular, Thales argued that divinities were present in all things and Anaximander described the Cosmos as a single organism fed by the pneuma.

Later Heraclitus of Ephesus described a world in constant change, an eternal "becoming". He believed that all the transformations of the world derive from the action of the opposites, and that each pair of opposites should be considered as a unit, to which he gave the name of Logos. From the opposites of Heraclitus, currents of thought originated which led to the separation between spirit and matter.

Parmenides of Elea, in sharp contrast to the eternal becoming of Heraclitus, called his fundamental principle the immutable Being, and argued that every change was pure illusion of the senses. From the thought of him, the idea of an indestructible substance was born at the basis of the properties of changing things.

From the contrast between eternal becoming and immutable Being, a thought developed that contemplated Being as present in certain invariable substances which, however, mixing and separating, gave rise to changes and different forms. Hence the concept of atom was born, as the smallest indivisible part of matter which, however, is the basis of everything despite all the apparent differences. Indeed today we know that all things are made of atoms.

The atomists, including Democritus, defined a clear separation between spirit and matter and imagined this as composed of "fundamental bricks", that is, completely passive and inert particles that moved in a vacuum, and whose movement could be due to external forces , forces of spiritual, not material origin.

This image then became predominant in Western thought, giving rise to the dualism between body and soul.

Aristotle, in his belief that problems concerning the soul were more important than those concerning the material world, created a pattern of thought that would be the basis of the conception of the universe throughout the West for almost two thousand years. Christianity also supported Aristotelian doctrines throughout the Middle Ages.

In the Renaissance the study of nature was faced using scientific methodologies for the first time, that is, applying controlled experiments to theoretical hypotheses. Galileo Galilei was the first to follow this path and is therefore considered the father of modern science.

The birth of modern science was accompanied by the development of philosophical thought which accentuated the "spirit-matter" dualism. In the eighth century, René Descartes described a nature founded on the fundamental separation between two distinct and independent realities, mind (res cogitans) and matter (res extensa). The Cartesian separation allowed the scientific world to consider matter as something inert and completely distinct from itself, and led to a representation of the world similar to an accumulation of different objects, objects that make up an immense lifeless machine.

The chariot of King Milinda.

The Milindapañha is a Buddhist text composed in the form of a philosophical dialogue, set in the city of Sāgalā, today's Siyālkoṭ. The dialogue takes place between the monk Nagasena and Milinda, historically identifiable with the Indo-Greek ruler Menander I who reigned in Punjab between 155 and 130 BC. about.

> "King Milinda went to where the monk Nagasena was and, after having greeted him politely and friendly, asked him:
> "How are you known and what is your name?"
> Nagasena replied, "I am known as Nagasena, O king, and by that name my brothers and parents address me."
> "You tell me that your brothers are used to referring to you as Nagasena. Now what exactly is Nagasena? Do you mean that the hair is Nagasena?"
> "I'm not saying that, great king."
> "Or body hair?"
> "Certainly not."
> "Maybe Nagasena is the nails, the teeth, the skin, the flesh, the nerves, the bones, the marrow? Are any or all of these things Nagasena? "
> To each of these questions the monk replied no.

“So the external form is Nagasena? Are sensations or perceptions or consciousness Nagasena? "

To each of these hypotheses he replied no.

“So, Nagasena is an empty and vain sound. Who is Nagasena that I see in front of me? Nagasen is a lie, a deception! "

Then the monk Nagasena said to King Milinda:

“You, Majesty, grew up in luxury. If you walk on stony or sandy ground, your feet will feel pain. So I ask you: How did you come here? Did you come on foot or in a chariot? "

“I didn't come on foot. I came in a chariot. "

“So if you came in a chariot, Majesty, explain to me what a chariot is. Are the horses the chariot? "

"I did not say this."

Is the chariot the axis of the wheels?"

"Of course not."

“But maybe the chariot is something different from what I have mentioned?"

Milinda answered again no.

“So," chariot "is an empty and vain sound. In which chariot would you have come? What you have said is a lie! Why are you saying false things??”

Then Nagasena called the Greeks and the monks as witnesses, saying:

“King Milinda said he came in a chariot. But when I asked him to explain what a chariot is, he was unable to do it. "

The Greeks and the monks applauded, and said to the king:

"Now, Majesty, answer if you are able!"

Milinda the king replied to Nagasena, saying:

> "I have not told any untruths, venerable sir. The ensemble consisting of the horses, the axle, the wheels, the frame, the ropes, the yoke, the spokes and the goad, indicates the common designation of chariot. "
>
> "Very good! - Nagasena said. You understand exactly the meaning of chariot. Likewise, it is the ensemble of hair, nails, teeth, perceptions, feelings and conscience that indicates Nagasena.."

So what is the chariot actually? Materialists would say that there is no chariot without the individual parts. The parts are the whole, and the chariot exists only as an accessory fact, an epiphenomenon whose presence or absence does not affect the existence of the parts. The fact that the parts, as a whole, go to constitute a chariot, is completely coincidental.

In a more spiritual view we would say that the single parts acquire a meaning only when, combined with intelligence, they give shape to the chariot, to Nagasena himself, and to every other thing existing in the universe. However, each logically assembled set of parts requires an intelligent design that goes far beyond the inertia of the materials that compose it.

The separation of Descartes.

The mechanical view of the world was endorsed by Isaac Newton, who used it to build his mechanistic science by laying the foundations of classical physics. The Newtonian view dominated the world from the mid-17th to the end of the 19th century.

Descartes' philosophy was not only important for the development of classical physics, but it had a huge impact on Western thought.

His famous phrase "Cogito ergo sum" (I think therefore I am) made man identify more with his own mind than with his body. As a consequence, man is aware of himself as an "other" who lives within his own body. The mind, separated from the body, receives the task of controlling it; this causes a conflict between will and instincts.

This separation from one's body is extended to the outside world, seen as a set of separate objects and events. The natural environment is made up of parts that can be grouped together in groups of interest (trees, stars, minerals or, indeed, the chariot of King Milinda).

Society is also divided into races, nations, peoples, religions and political groups, and our inner circle is divided between "my condominium, my relatives, my friends, workmates, those who have the same hobby , the patrons of my supermarket ". These divisions are at the root of all social and cultural crises and have caused that disorder and that unjust distribution of resources that often results in violence and war.

Cartesian thinking has brought benefits and harm; he must be credited with having introduced separation on a scientific level, while as regards social evils it is evident that they existed well before Descartes, to whom therefore no blame should be attributed but rather the clarity of having highlighted the separation between spirit and matter as the origin of such problems.

It is interesting to note how modern science, born from the Cartesian separation and from Newton's mechanistic conception, is rethinking these concepts and is returning to the idea expressed by the first Greek philosophies, that of a cosmos based on the cooperation of spirit and matter.

Why is there something instead of nothing?

Science often questions "how" things happen. He also asks "why" they occur, but with much less frequency. Indeed, deterministic beliefs presuppose that the "why" is taken for granted and can be logically found in a succession of causes and effects. "Something" causes "something else", and it is enough to reconstruct all the steps to get to understand the origin of the phenomenon. Therefore, the "why" is implicit in the "how", that is, there is no why.

This discourse becomes impractical when we talk about the first birth of the universe because, if we consider the moment before its existence, at that moment no cause was operating that started the chain of effects. Therefore, at this level both questions, about how and why, remain unanswered.

The sciences that try to solve this problem are cosmology and physics. The study of the cosmos has shown that it is in a phase of expansion, because it continues to extend into space. Starting from the current conditions and simulating a backward path, that is a contraction, we arrive at the point where the density of matter and its temperature become infinite, an impossible situation. Therefore, it can be assumed that the universe began at a specific time, placed between the current moment and a moment in the past temporally located before the conditions, receding again, became impossible.

There would have been a precise moment when the expansion would have begun and the universe would have "born". This, however, can provide us with information on how, but it does not explain why.

There are many cultural and religious traditions that attempt to provide answers. Bible scholars have long debated whether creation utilized pre-existing inert matter or took place out of nowhere. Both ancient Jewish exegesis and Philo, in the Hellenistic tradition, agree in supporting the creatio ex-nihilo (from nothing), a hypothesis still supported today by Christianity.

In many other cases, a universe is proposed that is remodeled starting from an initial situation of chaos, or from an unstructured void.

Sometimes the universe can spring from a primordial uterus, or it can be born from a pre-existing egg.

There are also traditions according to which the universe would have always existed, according to a conception of time inspired by agricultural civilizations, whereby everything would be subject to a perpetual cycle of death and rebirth.

In most of the hypotheses considered, excluding the biblical one, the pre-existence of a matter is taken for granted, albeit in the form of chaos, uterus or egg, and one does not ask where it comes from. However, all traditions do not explain how the transition process from nothing to something could have occurred, except by a divine creative act.

It all starts with the Big Bang.

The Big Bang is a cosmological model based on the idea that the universe began expanding at very high speed at a precisely definable time in the past and that this process continues to this day.

It is the predominant idea in the current scientific community, as it has had confirmations based on astronomical evidence and observations.

In particular, the fact that light elements such as hydrogen and helium are particularly abundant in accordance with the predicted values plays in favor of the Big Bang theory; another confirmation comes from the existence of the cosmic background radiation, with a coherent spectrum with respect to expectations. These and other experimental acquisitions have convinced most scientists that a Big Bang-like event actually took place about 14 billion years ago.

When it was proposed, the acceptance of the idea was neither taken for granted nor quiet. The Big Bang presupposed the creation of all energy and all matter (which is actually a different form of energy) in an initial act, more or less as Descartes and Newton argued and as argued by those who saw in the appearance of universe the creative intervention of God.

This circumstance produced the result that the scientific community was very reluctant to accept the theory. Instead the Catholic Church, with Pope Pius XII, fully accepted it in a speech he gave on Thursday 22 November 1951 at the Pontifical Academy of Sciences:

> "Two questions come to mind so spontaneously: is science able to say when this powerful beginning of the cosmos took place? And what was the initial, primitive state of the universe? The most excellent experts in the physics of the atom, in collaboration with astronomers

> and astrophysicists, have tried to shed light on these two arduous, but extremely interesting problems. (...)
>
> It is undeniable that a mind enlightened and enriched by modern scientific knowledge, which serenely evaluates this problem, is led to break the circle of a completely independent and autochthonous matter, either because it is uncreated, or because it is created by itself, and to go back to a Spirit. Creator. With the same clear and critical gaze, with which it examines and judges the facts, an enlightened mind glimpses and recognizes the work of creative omnipotence, whose virtue, stirred by the powerful "Fiat" pronounced billions of years ago by the Creator Spirit, unfolded in the universe ,. With a gesture of love, an exuberant matter of energy was called into existence. It really seems that today's science, suddenly going back millions of centuries, has managed to bear witness to that primordial "Fiat lux", when a sea of light and radiation burst out of nowhere with matter, while the particles of chemical elements they split up and reunited in millions of galaxies. " (Pius XII)

The Big Bang theory contained theological implications for which it was rejected and opposed, sometimes only because the Church supported it. Some renowned physicists advanced the hypothesis that this theory was part of a plot hatched by the Church itself. The astronomer Fred Hoyle condemned the Big Bang calling it a model built on Judeo-Christian foundations and formulated the alternative hypothesis of the stationary universe, where the creation of new matter occurs continuously; this theory, however, has been abandoned for some time because it would have predicted the birth of moving galaxies among those already existing, something not verified: there are no young galaxies positioned among the old

Figure 27- Niels Bohr, Danish physicist and initiator of the Copenhagen School. In the years following 1920 he perfected the classical model of the atom and was awarded the Nobel Prize in 1922.

ones. The younger galaxies, that is, those born earlier, are the farthest ones in accordance with the expansion of the universe that continues after an initial creative act.

Today the Big Bang theory is accepted by the absolute majority of the scientific community, without obviously accepting the idea that the initial creative moment originated from a divine will.

The atom and its classical representation.

The atom (from the Greek atomos, indivisible) is a basic structure of matter; in fact, all matter is made up of elementary units called atoms. This applies to matter in solid, liquid and gaseous states. It should be noted that this statement is difficult to connect to the plasma state, where the particularly high values of pressure and temperature prevent the formation of atoms. The atom was considered the smallest and indivisible unit of matter until, in 1897, Joseph Thompson announced the discovery of the electron, proving that the atom is actually divisible.

Today we know that atoms are made up of subatomic elements such as protons (positively charged), neutrons (uncharged) and electrons (negatively charged). Atoms form molecules by aggregating at higher levels.

The last classical model of the structure of the atom is the one developed in 1913 by Niels Bohr, which is still valid today without substantial variations.

According to this model, the atom consists of a nucleus, around which electrons rotate. The nucleus in turn is made up of two types of particles, protons and neutrons. Protons have a positive electric charge, while neutrons have a neutral electric charge. The electrons, on the other hand, have a negative electric charge and move in the outer space surrounding the nucleus, balancing its attraction with the energy of rotation, just as the rotation of the Moon balances the attraction of the Earth. Overall, the atom is electrically neutral since the number of protons (atomic number) and that of electrons is the same. In nature, atoms differ from each other in the number of protons present in the nucleus: the sum of protons and neutrons is called the mass number.

Currently the study of the atomic nucleus and its components is linked to the development of particle physics; in recent years it has been possible to define the structure of the nucleus more accurately, especially with the experiments conducted in accelerators. According to the Standard Model of particles, protons and neutrons are themselves made up of quarks.

To get an idea of the size of an atom these comparisons can be useful:

- the diameter of a human hair corresponds to about one million carbon atoms arranged in a row;
- if an orange became the size of planet Earth (about 12,700 km in diameter) its atoms would be approximately the size of the original orange (about 10 cm in diameter).

It is also important to note that the atom is essentially empty. In fact, if the nucleus of an atom had a diameter of 3-4 millimeters, its outermost electron would rotate at a distance of about one hundred meters.

The static representation of the atom just exposed is a valid working hypothesis but quantum physics is showing that in the subatomic level there are more probabilities than certainties. A certainty that has certainly been debunked is that the particles are "bits of matter", even if very small. Nothing more questionable.

Quantum physics.

Quantum physics is the theory that describes the behavior of matter and radiation, as well as the reciprocal interactions, in the very small length or energy scales, typical of atoms.

At the end of the 1800s classical physics appeared unable to describe the behavior of matter and electromagnetic radiation at the microscopic level, that is, at the dimensions equal to or smaller than those of an atom, or at energies relative to the internal bonds of atoms. Therefore, at the beginning of the 1900s it became evident that the birth of a new physics was necessary, completely different from the classical one developed up to then.

Thus was born the quantum theory, also called quantum physics or quantum mechanics. This branch of physics was initially introduced by Max Planck and is based on the fact that some quantities or quantities at the microscopic level, such as energy or "angular momentum", cannot vary linearly, but only by discrete values, called quanta.

In addition to the constituent particles of the main components of the atom, that is protons, neutrons and electrons, there are many others, and one could almost say that a new one is discovered every day. A particle that interests us in a particular way is the photon, that is the quantum of light: the particle that makes the stars bright and makes the sun shine.

The concept of how much is quite curious and was accepted with great difficulty by scientists of the 19th and 20th centuries.

At our macroscopic level we consider natural the concept that all quantities can take on a continuous set of values. For example, in addition to the kilo we also consider the quintal and the ton, as well as the hectogram, the gram, the milligram and so on, so it becomes natural that an object can weigh, for example, 7 kilos and 356 grams, i.e. 7365 grams.

Quantum mechanics, on the other hand, provides that the different quantities can assume only a determined set of values, multiples of a fundamental value that cannot be broken down. This value is called the "quantum of matter or energy".

Imagine a supermarket where you can buy loose fruit or 100-gram sticks of butter. It will be easy for you to put 1,365 kilograms of apples in the cart, while in the case of butter, you can only buy 100 or 200 or 300 grams, depending on the number of blocks you put in the cart. The world of particles predicts that energy and matter are not available in bulk but only in the form of blocks, or bricks, which however are called quanta.

Therefore, in the world of particles, energy and matter, in addition to not being able to be reduced to lower values of the "quantum", are available only as multiples of the quantum itself; any of

Figure 28 - Albert Einstein, probably the best known physicist and mathematician of our times. He challenged the quantum theories of Niels Bohr and the Copenhagen School by devising the EPR experiment to demonstrate its inconsistencies. The subsequent evolutions of the research instead led to their confirmation.

their values must correspond to integer quantities of quanta. As is easily understood, the quantum has an extremely low value.

In the universe, everything is made up of how many, even the book you are reading. The reason why you do not feel its "graininess" is that the quanta are so small that our senses cannot perceive them, just as we do not perceive individual atoms or even single molecules in a distinct way.

One of the first cases studied is the quantization of energy: for example, the electrons of an atom can only have certain energy values, so they can rotate around the nucleus only on certain orbits, not on any orbit. Similarly, photons are the quanta of the electromagnetic field: they are non-divisible packets of light.

The quantized atom.

The concept of "quantum" is very important for the functioning of matter in the extremely small, that is, in atoms, but consequently also in the entire universe. To understand it better, I can propose a similarity with some device commonly used in everyday life. For example, let's imagine a traditional-type gas cooker, equipped with hobs arranged on one surface and a built-in oven. The kitchen controls are aligned on a front panel and consist of dispensing knobs that allow the gas to be sent to the respective burners.

Most of the knobs can be rotated linearly from left to right, in order to block the gas when it is completely turned to the left, or to send it to the relative stove in increasing quantities, as it is turned to the right. until reaching the limit switch, i.e. the maximum allowed. Let's say that the knob allows a linear (or continuous) increase in the amount of gas.

One of the knobs, however, in particular that of the oven, works in a different way: it cannot rotate continuously, but only in jerks, allowing the flow of gas in certain quantities, suitable for obtaining the heat signaled at each click with values expressed in degrees centigrade: 120 ° C, 180 ° C, 270 ° C or similar values. In this case the gas cannot increase continuously, but rather discreetly: that is, only for fixed and determined quantities.

In truth, even the knobs on the cooker work discreetly. The problem is that the variations between one point and another are so small that they give us the impression of a linear increase or decrease. Put simply, our senses can't catch the variation shots, because they're too small.

In the average level of the dimensions of matter, which is the one in which we live and our senses operate, everything seems to us to increase (or decrease) in a linear way. The diffusion of light at dawn or dusk, the sound of a car approaching or moving away, a fly buzzing around us. On that fly we would say everything, but we certainly wouldn't say that it flies in jerks.

This once again gives us the measure of how deceptive our senses are because, in reality, nothing moves linearly but everything does it in a discrete way, that is, by quantity of energy, or quanta.

The difference between our macroscopic level and the microscopic level of particles is that we rather crudely evaluate movements and energies as linear.

Practically, according to a convention used in physics, a material body at the macroscopic level can be studied as a continuous body, since the very high number, the cohesion and the interdependence between the particles make any granularity disappear.

Instead, in the infinitely small it is no longer possible to ignore this feature because the dimensions of subatomic bodies are sensitive to quantum graininess.

In conclusion, in quantum mechanics the quantum (from the Latin quantum which means quantity) is a discrete and indivisible quantity. By extension the term is sometimes used as a synonym for "particle".

Quantum mechanics was mainly initiated by Niels Bohr, Louis-Victor de Broglie, Paul Dirac, Albert Einstein, Enrico Fermi, Wolfgang Pauli, Erwin Schrödinger, Richard Feynman and Werner Heisenberg.

The intuition of the quantization of matter was born in 1900 from the work of Max Planck, who adopted this convention to confirm his theory relating to the electromagnetic emission of the black body.

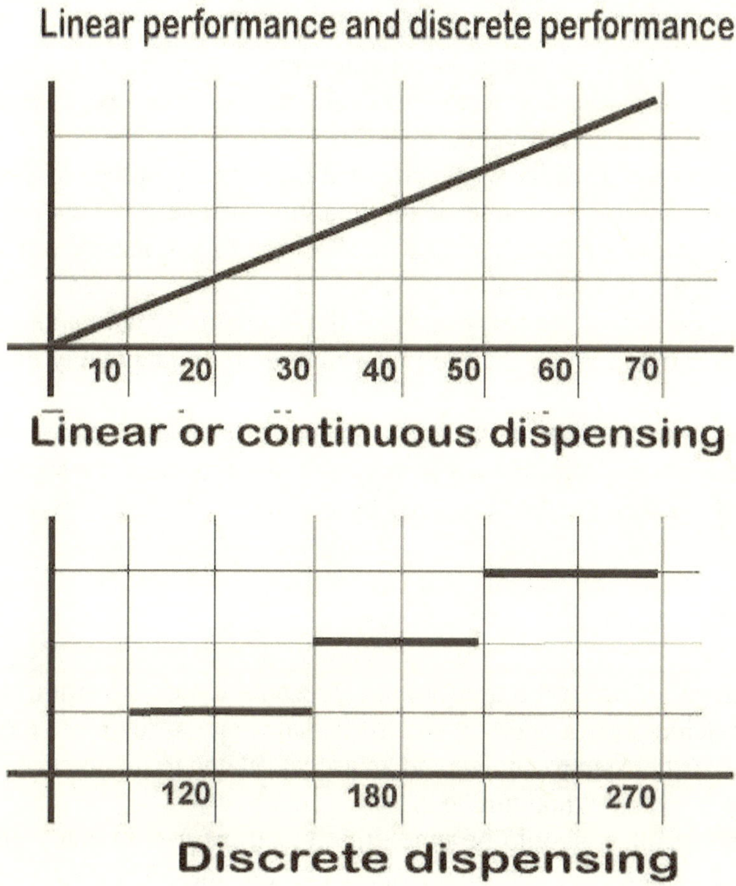

Figure 29 Linear course and discrete course.
1 Linear or continuous delivery.
2 Discharge with discrete values

- The linear trend assumes an (almost) infinite number of intermediate levels. The discrete trend, on the other hand, includes only a few planned increases.

In practice, Planck noted that his hypothesis about black bodies would only work in the presence of quantized energy. However, Planck was very reluctant to consider this possibility as real. Although he was the first to theorize this reality, he always found it difficult to consider it valid.

This physical reality was confirmed only later by other scholars who determined the value of the quantum of energy. This value took the name of Planck's constant, because they honestly wanted to recognize the work of those who had guessed and used it first. Planck's constant, or magnitude of the "quantum", is denoted by the letter h and is 6.6 × 10−34Js, which is an unimaginably small value.

The hypothesis of the granular structure of mechanical energy found full and definitive confirmation in the theory of the specific heat of solids formulated in 1907 by Einstein.

The quantum leap.

Still referring to the traditional representation of the atom, that is, a nucleus surrounded by electrons that rotate around it, we can take a further step to understand an aspect related to its functioning which is called a quantum leap.

First of all, it should be noted that electrons cannot rotate in a regime of anarchy, but must respect certain rules. The first is the Pauli Exclusion Principle, which applies to fermions, a class of particles that includes protons, neutrons and electrons.

This principle starts from the concept that around an atom there are only a few "allowed zones", and that two identical fermions cannot simultaneously occupy the same zone (or quantum state).

Therefore, an orbit can be occupied by at most two electrons, provided that they have opposite spin.

The second condition is that each orbit cannot be randomly distant from the nucleus.

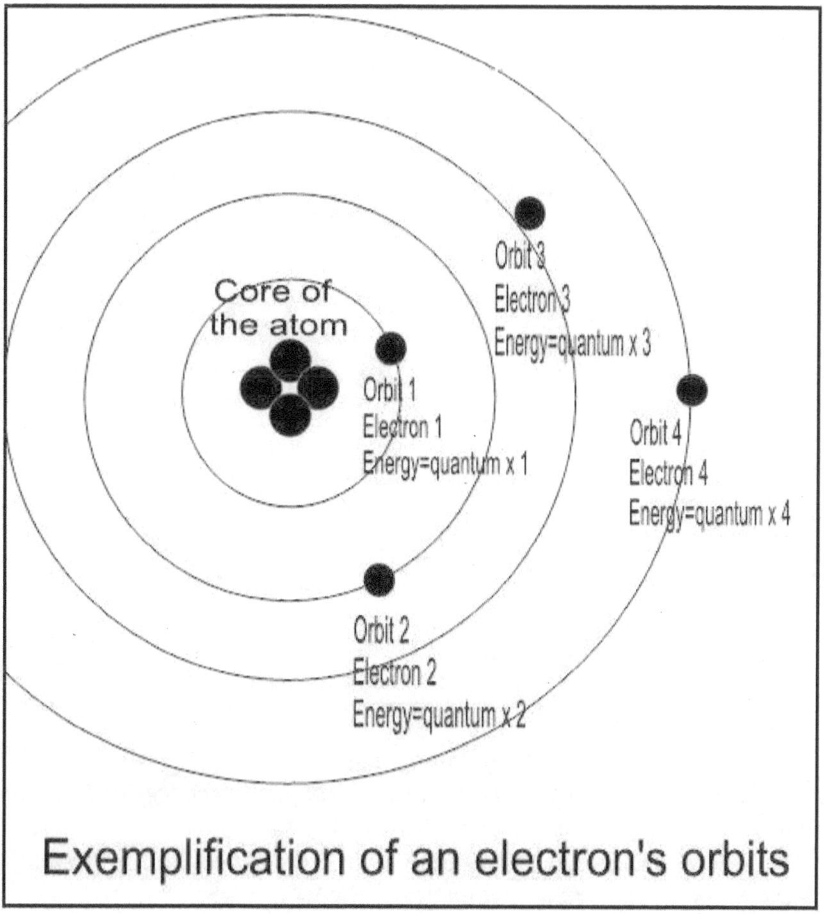

Figure 30 - Example illustration of an atom with four electrons and four protons. Each electron rotates on a distinct orbit and has an amount of energy that can be measured in quanta.

1 The electron on orbit 1 has 1 quantum of energy.
2 The electron on orbit 2 has 2 quantum of energy.
3 The electron on orbit 3 has 3 quantum of energy.
4 The electron on orbit 4 has 4 quantum of energy.
5 Nucleus of the atom.

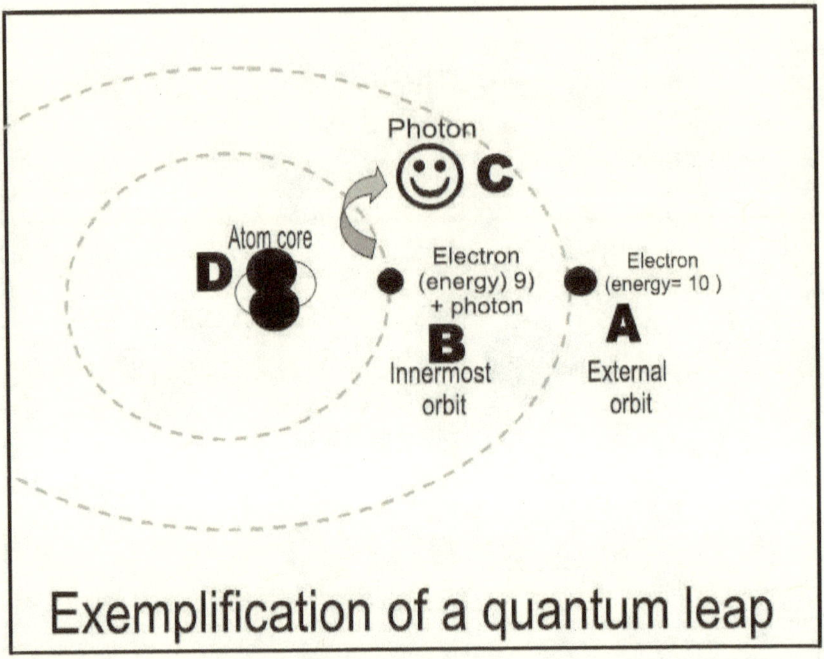

Figure 31 - Example of a quantum leap.

A the electron A placed on an external orbit has an energy of 10 quantum.

B The electron jumps to an inner orbit. Here it needs a lower energy (9 quantum) so it gives up a quantum.

C The quantum released becomes a photon.

D Nucleus of the atom.

According to the atomic model proposed by Niels Bohr, the electron rotates around the nucleus describing a circular orbit; each electron, which has a negative charge, corresponds to a very precise orbit, as it must move at a speed sufficient to counteract the attractive electric force that the positively charged nucleus exerts on it. The rotation prevents the electrons from being sucked by the nucleus, but it must take place with an energy proportional to their distance.

The farther an electron is from its nucleus, the more energy the electron needs; but it must be a well-balanced energy. The right amount of energy must prevent the electron from falling into the nucleus or being blown away, outside the atom's area of influence.

Since we know that the energy at the level of electrons is measured in quanta, then the orbits can be placed only at the distance of a quantum, or multiples of a quantum.

No electron will ever rotate on an orbit that requires a non-integer value of energy measured in quanta.

Therefore, between an electron that rotates on an internal orbit (let's call it E-Internal) and another that rotates on an outer orbit (let's call it E-External) there is a difference of energy: E-External needs more energies than E-internal.

As an example, we can say that if the inner electron needs one quantum of energy, the outer electron requires two to stay on its orbit. There can be no orbits that require a quantum and a half, because the quantum is the smallest possible measure.

In the midst of so much order we must consider that electrons are rather undisciplined and sometimes, for reasons known only to them, they like to jump from an outermost to an innermost orbit. So, if an electron that has two quanta of energy, wanted to jump onto an inner orbit, (which would require only one quantum of energy), it would have to give up one of its quanta. What happens to this released energy? It takes the form of a photon as the electron makes what is called a "quantum leap".

Figure 31 represents a quantum leap. At point A, the electron, placed on any orbit, has an energy equal to 10 quanta (an example value). At point B the electron jumped to an inner orbit. Now the energy required is equal to 9 quanta, and the unnecessary quantum has been transferred in the form of a photon: it has become a quantum of light.

As we will see later, physicists have exploited this property by stimulating electrons to make a double quantum leap to produce two twin photons or related photons; studying their behavior led to the discovery of the phenomenon called entanglement.

The wave-particle duality.

The fundamental characteristic of quantum physics is that of being able to attribute to radiation and matter both the form of a wave phenomenon and that of a particle entity. This is in contrast to classical physics where, for example, light is described exclusively as a wave and the electron exclusively as a particle.

This completely unexpected property is called wave-particle duality. Not having considered it constituted the reason for the failure of all the classical theories previously developed on the functioning of the atom. A defining feature of quantum physics is the principle of complementarity, according to which the two aspects, wave and particle, cannot be observed simultaneously during the same experiment. This concept was formalized in Heisenberg's uncertainty principle. Heisenberg states that it is not possible to measure simultaneously and with extreme accuracy the properties that define the state of an elementary particle. If, for example, we could determine its position with absolute precision, we would have maximum uncertainty about its speed.

The advent of quantum physics has caused bewilderment and disturbance in the scientific community which saw the foundations of its conception of the world questioned and revolutionized.

Einstein himself, when he became aware of the new reality of quantum physics, called it "a sinister action at a distance".

As reported in the book "Albert Einstein scientist and philosopher", in his diary egli he wrote as follows:

> "All my attempts to adapt the theoretical foundations of physics to these acquisitions failed. It was as if the earth was missing under our feet, and there was nowhere to be seen a fixed point on which to build ".

The discoveries of quantum physics made it necessary to revisit fundamental concepts such as matter, time, space and cause-effect relationships, all indispensable for understanding the world as we perceive it. Surely the need to modify them produced profound cultural crises in the scientific community. Even today it is not possible to fully define what the exact vision of physical reality is, if not examining it separately on different levels: that of the medium dimensions, that is our sensitive level, and that of the subatomic dimensions.

The double slit experiment.

The achievements of modern physics have provoked a significant change in our scientific image of the world: they have shaken the absolute validity of natural laws and have transformed it into a relative validity. Natural laws are statistical truths, that is, they are, as it were, entirely valid only when it comes of macrophysical quantities. In the sphere of minimal magnitudes, instead, predicting what will happen becomes uncertain and impossible, since very small quantities no longer behave in accordance with known natural laws.

(Carl Gustav Jung)

One of the best known experiments in quantum physics (probably the best known of all) will help us better understand the nature of elementary particles. The experiment bases its importance on its ancient origin: the first version was made over two centuries ago, in 1801, by Thomas Young. Subsequently, the experiment maintained its dignity which made it current in many subsequent re-propositions, updated as technology made more current versions possible. At the moment it represents one of the main keys to understanding quantum physics and related phenomena, including entanglement. I will therefore try to illustrate this experiment in such a way as to make it easily assimilable to the non-specialist reader and avoiding mathematical formulas.

The experiment is not conceptually complicated and I will propose an example that, I believe, can be understood by anyone, as a boy, has enjoyed throwing stones at targets of all kinds, or throwing them into the water to see the formation of typical concentric circles. .

Raise your hand if you have never thrown a stone: and whoever has done so, prepare to throw others, even if he is over the age.

The game consists in fixing a sturdy barrier in front of us, in which we will have drilled a hole or slit. The slit is large enough to allow a stone thrown in its direction to pass. Just behind the barrier we place a thin sheet so that, when hit by a stone, it dents it, leaving the impact mark.

At this point the experiment can be performed, as shown in figure 34. Let's put ourselves a few meters away and throw as many stones as we want, trying to get them to pass through the crack. We will only consider the stones that pass through the crack and hit the receiver screen, and we will ignore those that do not hit and are repelled by the barrier.

Each stone hitting the receiver screen produces a single dent (fig. 34-1).

If we open a second slit in the barrier (fig. 34-2), the successful launches will pass through one or the other slit, and will produce dents on the receiver screen in correspondence with the slit crossed.

Figure 32 Thomas Young (Milverton, June 13, 1773 - London, May 10, 1829) was a British scientist, famous for his research on light and the mechanics of solids, and for his contributions to physiology and Egyptology.

Figure 33. Werner Karl Heisenberg (Würzburg, 5 December 1901 - Munich, 1 February 1976) was a German physicist. He was one of the main architects of quantum mechanics, he is known for having exposed the uncertainty principle in 1927. He received the Nobel Prize in Physics in 1932.

In summary, in this part of the experiment we can understand that the barrier was crossed by single solid bullets simply by observing the dents produced on the receiver screen. We can now move on to the second part of our experiment. We will immerse the device, ie the barrier with a slit and the receiver screen, in a vessel containing water, so that the slit is semi-submerged, as in Figure 35

This time we will not throw our stone directly towards the crack, but into the water, some distance in front of the barrier. The stone will produce the typical water circles that we all know; these will propagate towards the barrier, cross the crack and re-emerge on the opposite side, forming new propagation waves which will be blocked by the receiver screen. Now imagine opening a second crack and throwing another stone, as in figure 35.

The water circles created by the single stone, propagating in front of the barrier, will cross both cracks. Emerging on the back of the two-hole barrier they will produce two new circles of water. These circles, spreading further, will interfere with one another by adding up.

From what has been said so far we can draw some simple conclusions:

- If we throw bullets towards the double slit barrier, the projectiles go through one or the other slit. We can verify this by observing the dents on the receiver screen.
- If we throw a stone into the water towards the barrier with two slits, we produce a group of water circles. These circles pass through both slits, and produce two new sets of water circles on the back of the reef. These two groups will join and produce an interference pattern.

So we saw the difference between throwing a stone and throwing a wave: throwing stones produces dents, while instead, throwing propagation waves produces interference figures.

When the scientist wanted to understand if light is made of particles or waves, someone imagined throwing "flashes of light" towards a device like the one just described. The experiment tended to verify what effects would be produced on the receiver screen, "dents" or "interference".

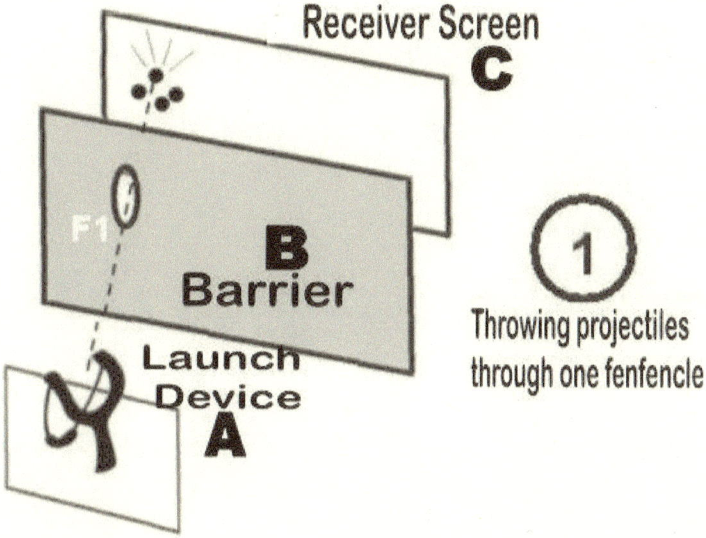
Receiver Screen
C
F1
B
Barrier
1
Throwing projectiles
through one fenfencle
Launch
Device
A

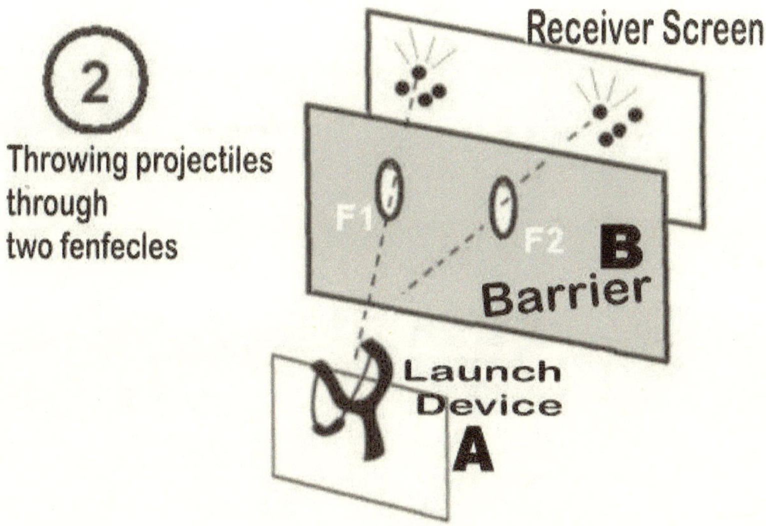
Receiver Screen
2
Throwing projectiles
through
two fenfecles
F1
F2
B
Barrier
Launch
Device
A

Figure 34 - Simulation of the double slit experiment. 1) Throwing projectiles through a rift. 2) Launch through two slits.

1) The launch occurs through a barrier with a slit.

A - Launch device

B - barrier

C - Receiver screen

2) The launch occurs through a barrier with two slits.

A - Launch device

B - Barrier

C - Receiver screen.

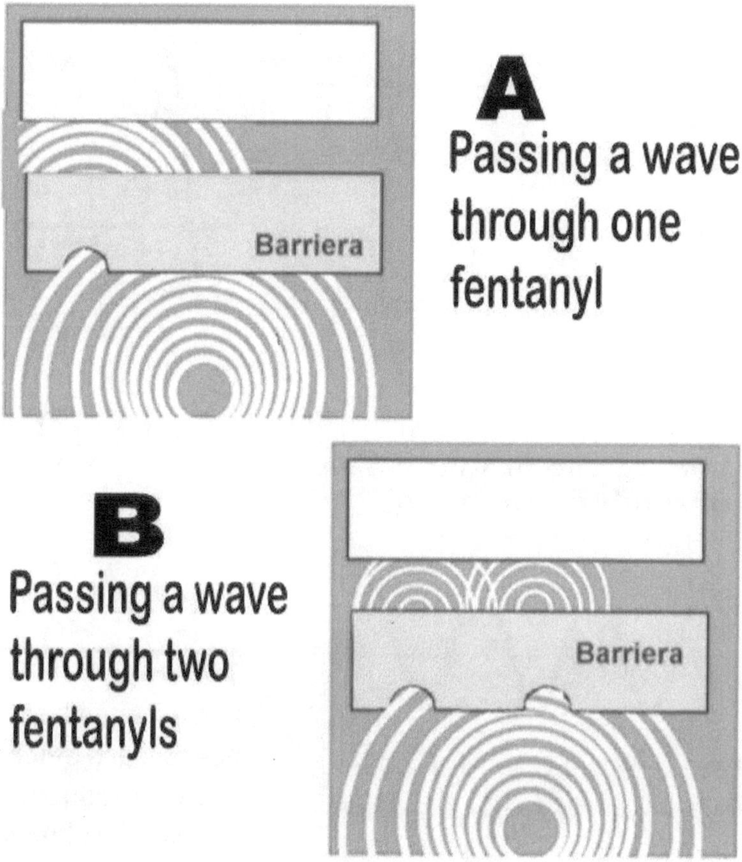

Figure 35 - Passing through two slits, the wave produces two half-waves which, intersecting, add up.
A) Passage of the wave through a slit.
B) Overlap effect that occurs when the wave crosses two slits.

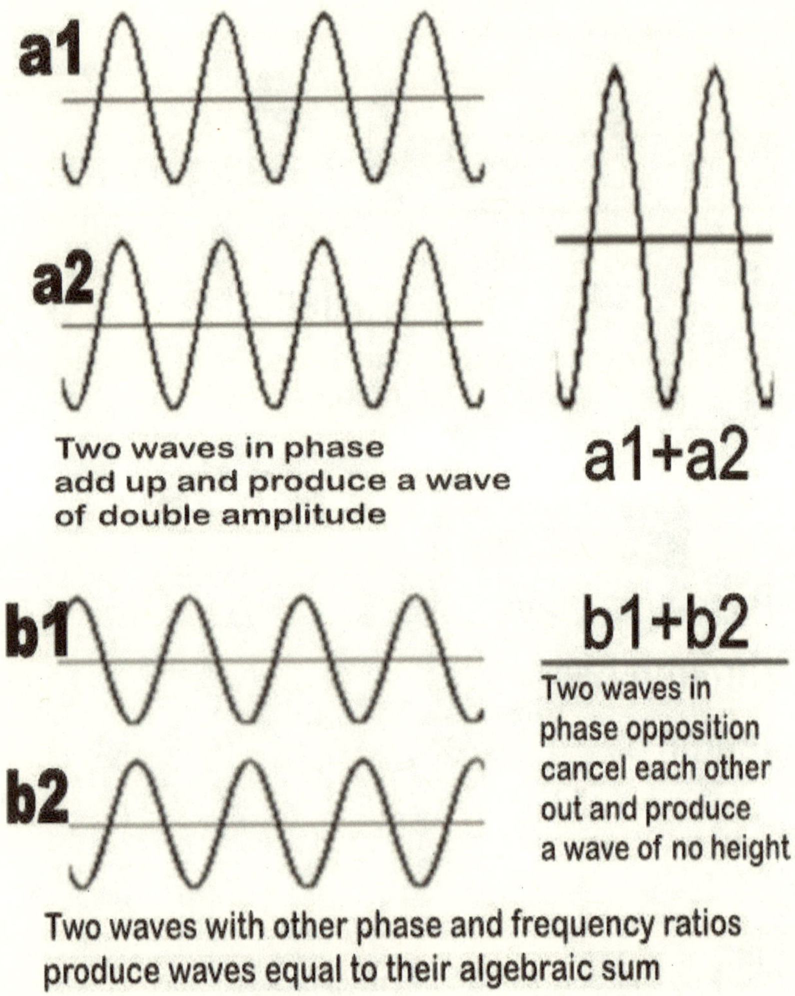
a1
a2
a1+a2
Two waves in phase
add up and produce a wave
of double amplitude
b1
b2
b1+b2
Two waves in
phase opposition
cancel each other
out and produce
a wave of no height
Two waves with other phase and frequency ratios
produce waves equal to their algebraic sum

Figure 36 - Sum of waves with different amplitudes, frequencies and phases

A) The two waves a1-a2 are in phase. Their amplitude adds up and produces the a1 + a2 wave of double amplitude.

B) The two waves b1-b2 are in phase opposition. Their amplitude is added and produces a negative result, the b1 + b2 wave of zero amplitude.

If two waves are not exactly in phase or in phase opposition they produce a wave equal to their algebraic sum.

Thomas Young

Thomas Young was a British scientist, well known for his research on light and the mechanics of solids, but also for his contributions to Egyptology. He was born in Milverton on June 13, 1773 and died in London in 1829.

He was the first of ten siblings in a Quaker family. He is described as a child prodigy, and he certainly was. In fact, at the age of two he already knew how to read. By the age of six he had already learned Latin and read the entire Bible twice. He soon learned other languages and at the age of fourteen he knew, in addition to Greek, French, Italian and Hebrew, and many oriental languages such as Chaldean, Syriac, Samaritan, Arabic, Persian and Turkish.

Young graduated with a degree in physics in 1796 in Göttingen, and in 1799 he began working as a physician in a study on Welbeck Street in London. A plaque on the door still remembers that collaboration today. Meanwhile, he published his essays anonymously because he feared repercussions on his reputation as a doctor.

In 1802 Young joined the Council of the Royal Society, of which he had been a member since 1794.

In physics Young is well known for his work in the field of optics, as he was the author of a research on wave-particle dualism; the discovery of the phenomenon of interference in light is also due to him. In fact, in 1801 he devised the double slit experiment.

The diatribe about the nature of light began with Newton considering it to be of a corpuscular nature. Later this hypothesis was contested, so in 1801 Young designed and carried out his famous experiment. He sent beams of light towards a barrier where two holes had been drilled. On the receiver screen placed behind the barrier, he obtained an interference pattern. Hence, Young concluded that the barrier had been crossed by waves rather than by corpuscles, and confirmed the wave nature of light. But is light really a wave?

Evolution of the double slit experiment.

At the beginning of the twentieth century, doubts began to arise related to Planck's studies on black bodies and on some aspects highlighted by Einstein regarding the behavior of the photon, which very much resembled a particle if it was studied in the photoelectric effect.

In 1916 the American Robert Millikan hypothesized the existence of the photon as a particle, as was later confirmed by the discovery of the Compton effect in 1922. This created a big problem in physics, as light seemed to possess a mysterious dualism that made it appear as a wave or as a particle in experiments of different types.

In 1924, the French physicist and mathematician Louis de Broglie formulated the hypothesis that not only photons, but all material bodies, and particularly all particles, also possessed the typical physical properties of waves. This hypothesis soon found experimental confirmation and gave a fundamental impetus to the development of quantum mechanics. Experimental proof of wave-particle duality was obtained in 1927 by Clinton Davisson and Lester Germer, at Bell Laboratories in Murray Hill, in the state of New Jersey of the United States of America. These scientists fired an electron beam at a nickel crystal, and observed diffraction patterns. This made it possible to use particle beams to perform interference experiments with two slits, just as Young had done.

Modern version of the double slit experiment.

The double slit experiment was performed for the first time in modern form using electrons in 1961, by Claus Jönsson at the University of Tübingen, in Baden-Württemberg in Germany. It was then repeated in 1974 at the LAMEL Institute of the CNR in Bologna by Pier Giorgio Merli of the LAMEL and by Gian Franco Missiroli and Giulio Pozzi, both of the Physics Department of the University of Bologna.

The LAMEL Institute had a latest generation microscope, a Siemens Eliskop 101, which was calibrated and adapted to operate in interference electron microscopy.

The microscope was used both as an interferometer and as a source of electrons that were passed one at a time through a double electronic prism to be then sent to a photographic plate.

Description of the experiment with the launch of a photon.

Now we want to consider in detail the experiment in its classic modern version, based on the launch of single photons, or quanta of light. So let's imagine a device capable of launching single photons one at a time, a barrier with one or two slits and a rear screen consisting of a dark room containing a photographic film that will be impressed by the photons received, as these are sources of light.

The device in the configuration of a single slit is shown in Figure 37. Photons which are rejected by the barrier are not considered; those that cross the slit mark a dot on the photographic film, a clear sign that they have behaved like particles. So far nothing extraordinary; photons behave as our stones did, and the experiment suggests that photons are particles.

Figure 38 illustrates the launch of a single photon towards a barrier with two slits. The question mark symbolizes the question: which of the two slits will be traversed?

The answer is evident from Figure 39: the single photon has passed through both slits.

In fact, in front of a single slit the photon behaves like a particle, while in front of two slits it behaves like a wave and passes through both of them.

In the first case, the confirmation is given by the single luminous dots that appear on the photographic film. In the second case, dark and light bars appear on this, that is, the interference figure given by the wave that has crossed both slits, creating two outgoing waves. Beyond the barrier the two waves interfere and algebraically add to each other. The dark or light bars represent the result (fig. 39).

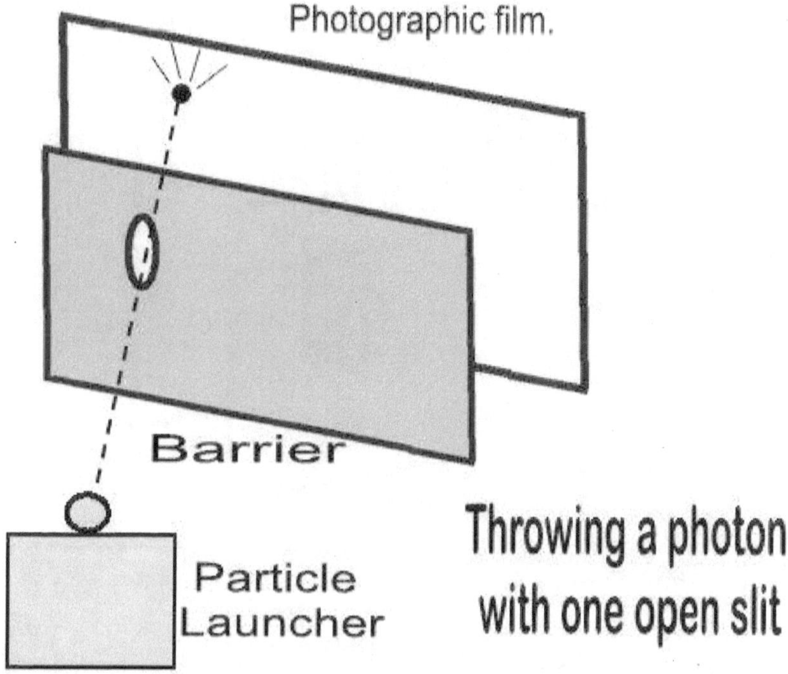

Figure 37 - Launch of a single photon towards a barrier with a slit. The photon crosses the slit and cuts a point on the photographic plate behind it. Hence, it behaves like a particle.

A) Launch device

B) Barrier with a slit

C) Photographic film.

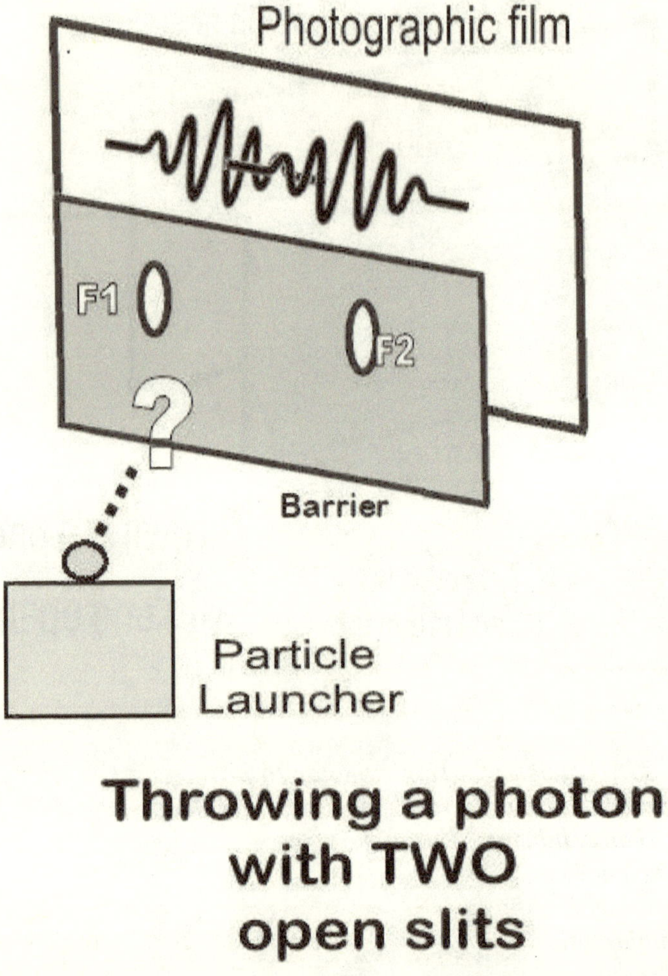

Figure 38 - If we were to launch a single photon towards a barrier with two slits, this, as a particle, would pass through only one while, as a wave, it would cross both.

A) Launch device

B) Barrier with two slits

C) Photographic film.

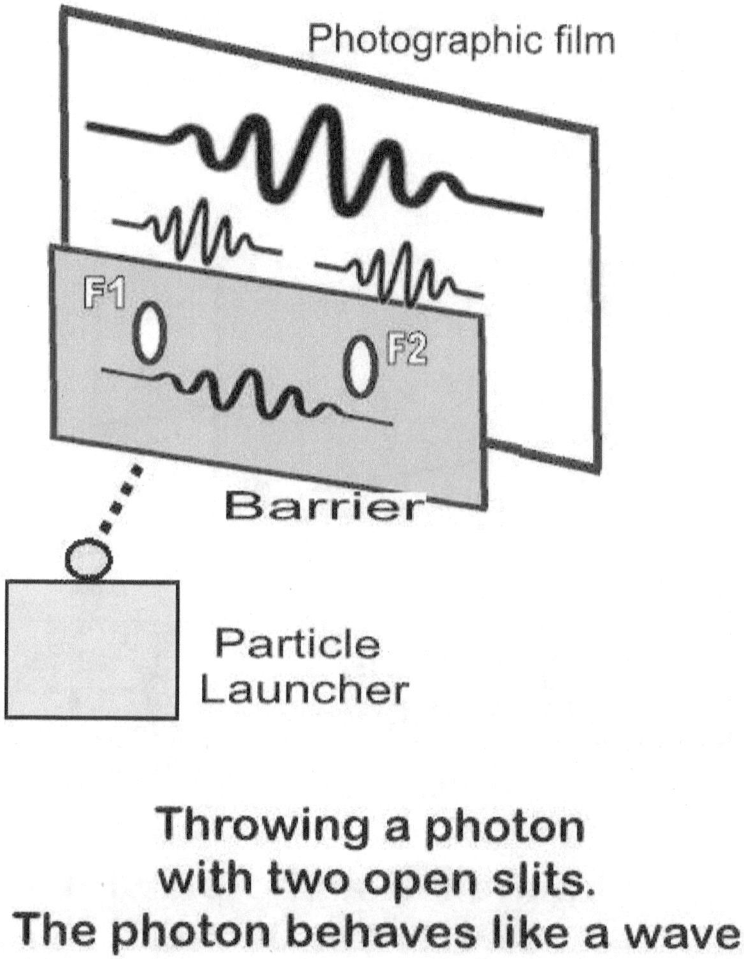

Figura 1 - Nei fatti il singolo fotone, se ha di fronte una sola fenditura, si comporta da particella; se ne ha davanti due, si comporta da onda.

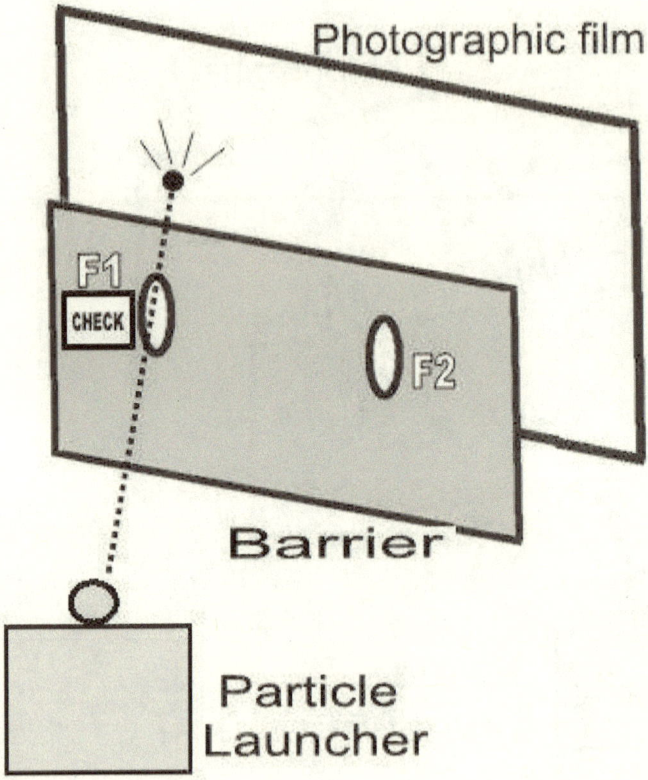

Launch of a photon with
CONTROL DEVICE.
The photon behaves like a particle

Figure 40 - If we launch a photon against a barrier with two slits, one of which is equipped with a control device to verify the passage of photons, then our photon chooses precisely the controlled slit and behaves like a particle.

A) Launch device

B) Barrier with a slit

C) Photographic film

CHECK) Control device.

A first consideration that can be made is that photons (but not only, the property is common to all elementary particles) behave, depending on the case, as particles or as waves, in the sense that they are both particles and waves at the same time.

This makes it possible for a single photon to pass through both slits at the same time.

But the surprises aren't over yet. Let's imagine that the experimenter decides to verify the passage of the photon through the slit, starting from the principle that in any case the photon is a particle, and must pass somewhere.

Therefore, he places an instrument just beyond one of the two slits (or both) to verify which one is crossed by the photon (Figure 40).

Well, in this case, that is, when the passage of the photon is controlled by an instrument, something incredible happens, almost as if the photon refuses to cooperate. Although the open slits are two, the photon only passes through the one equipped with the instrument, and goes to impress a dot on the screen behind it. If both slits are monitored, it passes through only one and still behaves like a particle. But if you take off the tools, it goes back to acting like a wave.

In practice, the photon challenges the experimenter by showing him to be, for him, an unsolvable mystery.

Leon Max Lederman American physicist, writes in his book Quantum Physics for Poets:

> “It seems that a single electron, as if by magic, goes through the two slits at the same time if nobody observes it, while choosing a possible path if someone or something looks at it! This is possible because electrons are neither particles nor waves: they are another thing, completely new. Particles are "quantum states."

It must be said that the experimentation, initially limited to elementary particles, was then also performed on complex bodies. At the University of Vienna, Anton Zeilinger and his colleagues involved "bucky balls" in the experiment. These are molecules made up of sixty or more carbon atoms assembled to form a structure that resembles a geodesic dome; this arrangement was made famous by the American architect Richard Buckminster and was named after him. Obviously, a molecule comprising sixty atoms becomes a gigantic structure when compared to a single photon. Despite this, in the double slit experiment, using "bucky balls "instead of photons, the same results were obtained.

With this it has been proved that a molecule of matter can produce wave phenomena such as an energy particle. Therefore, these molecules of matter also have the gift of ubiquity, that is, passing through both slits, they have the ability to be simultaneously in two different places.

Does bilocation become plausible?

It will be possible in the future to perform the experiment on more complex bodies, composed of billions of molecules, how could a human body be? In this case, there would be scientific confirmation of controversial phenomena such as doubling, astral projection, out-of-body travel or bilocation, i.e. all out-of-body experiences, also known as OBE and sometimes OOBE (out of body experience). They are all experiences, the interpretation of which remains controversial, in which a person perceives to "go out" from his physical body, that is to project his conscience beyond the bodily boundaries; they belong to the field of parapsychological disciplines, which are not recognized by the scientific and academic world, but are present in many cultures, both Western and Eastern.

The phenomenon of bilocation has been part of the Vedic and Buddhist tradition for many centuries; this culture would be able to transmit, to subjects with particular inclinations, the philosophical and spiritual knowledge bases to allow them, through a long training, the ability of bilocation. The well-known Indian philosopher

Paramahansa Yogananda testifies in his books, in great detail, episodes concerning himself as well.

Robert Allan Monroe is an American writer, founder in 1978 of the homonymous Institute "for research on human consciousness". In his book "My travels out of the body" he describes the oriental disciplines and recommends preparatory techniques for initiation into the phenomenon. of bilocation.

In Christianity there are many hagiographic testimonies of bilocation concerning, among others, St. Anthony of Padua, St. Francis of Assisi, St. Alphonsus Maria de 'Liguori, St. Joseph of Copertino, St. Francis of Paola, St. John Bosco.

The precognitive abilities of particles.

From the double slit experiment we can draw conclusions that will leave even the most expert physicists astonished, and which probably relate more to the philosophy of physics than to physics itself.

Let's imagine for a moment that the photon has its own consciousness and is able to "decide" its behavior in the various situations in which it is involved.

In practice we ask ourselves: when a photon is launched towards the barrier, how does it know if it will have one or two open slits?

Why, in fact, the photon should know how many slits there are only after crossing the barrier. Apparently, however, the photon knows first. In fact, the photon needs to face the barrier in the form of a particle or in the form of a wave, before having crossed it.

Not only; the photon knows, before passing through a hole, if this is monitored by an instrument to verify its passage. The photon knows which of the two holes it is monitored before passing through. In the very brief moments of its existence between the moment it is launched and the moment it encounters the barrier. the photon acquires knowledge of what awaits it. Egi knows if he will find a crevice or two, and if they are monitored or not.

Since the photon has no mind or consciousness, we must ask ourselves where its ability to "know first" what awaits it in its immediate future comes from. The provocative title of this chapter, on the precognitive abilities of the photon, would suggest that these really exist.

Probably yes. The photon knows because it draws knowledge, indeed it is part of a universal wisdom, a knowledge that connects the whole cosmos. The photon knows because the whole cosmos knows, and this knowledge has no limits of time, space or size. The photon does not "know before", it just knows. He does not know due to the fact that the barrier is close in space, he would know even if it were at the outer limits of the universe. The photon knows why it is part of a universal cosmic wisdom placed beyond all limits and beyond any constraint determined by matter. His knowledge is information and intelligence on a pure psychic or spiritual level. That single photon, and every other particle in the universe, does not predict, it just knows.

Then another question arises: why do we human beings, who are basically made of particles, not possess the same knowledge?

Of course, all the particles that make up our body possess this knowledge, each one individually; we, considered as the aggregate of an endless number of particles, should possess an aggregate knowledge, the sum of the knowledge of all our particles.

There is a big difference between the single particle and the aggregate of particles that we are. First of all, particles are members of universal knowledge without any mediation. The particles are directly part of it. Instead our body, as an aggregate of particles, is affected by powerful intermediaries which, in fact, become separators.

These negative mediators collectively have the name of "I". Our consciousness and our personality perform the function of separators towards cosmic knowledge. Indeed, while particles are humble members who docilely adhere to this knowledge, we place ourselves at the center of the universe: everything around us, what happens around us, memories of the past and expectations of the future are absolutely centered on the our personal cosmos.

This makes us very little compliant and not very permeable with respect to a knowledge that is not at all centered on ourselves.

Considerations similar to these instinctively evoke the terrible cry of Lucifer cast out of the presence of God: "I will not serve you!". Perhaps the fallen angels are ourselves, since we really do not wish to serve anything other than our ego and this plunges us into an existential loneliness made up of selfishness and self-referentiality.

So too the biblical myth of Adam and Eve tells of an expulsion from Eden due to disobedience. God had planted an orchard that the couple could fully benefit from, with one exception: “From every tree in the garden you will be able to eat; but of the Tree of the Knowledge of Good and Evil you will not eat; because the day you eat it you will surely die ».

The knowledge of Good and Evil transforms Adam and Eve into subjects capable of judgment, but this ability, unfortunately, is vitiated by the ambition of wanting to be similar to God.

Despite this, there are many aspects that keep us connected, albeit faintly, to a "cosmic knowledge". Despite being foreclosed on, we possess crumbs of awareness of its existence. First of all we have the sense of a presence superior to ourselves, which we perceive and which we sometimes feel the need, as it could be for the memory of distant or missing parents. The shadow of an unknown parent to be embraced mystically swirls in our thoughts, intrigues and fascinates us. This is the feeling that many call "longing for God". It is a nostalgia that comes from far away, evoked by events that are often not part of our life experience; it is a nostalgia that draws on memories prior to our physical existence.

Then there are extrasensory perceptions, that is, those flashes of light that illuminate at times and make the path that unites us to cosmic knowledge passable. Our unconscious is able to connect with the collective unconscious and, albeit tortuously, flashes of knowledge flow from the universal consciousness to our individual consciousness. When this happens we "know and see", for a few brief moments, the things that, due to our condition as slaves chained to matter, we do not normally see. .

But the most concrete link with the wisdom of the cosmos is found in the synchronicities that crowd our existence, often in our indifference. While we have sunk into our abyss of fallen angels, a hand reaches out to us to lift us up from the pit and guide us towards a spiritual and cultural evolutionary redemption. SSynchronicities are proof that "Knowledge" does not consider us cursed forever and sees in us that evangelical prodigal son who, on his return, is not rejected by his father but welcomed with joy, to the point that his father prepares a table for him with the fatted calf.

The atom and its most current representation.

In the classical representation of the atom, an electron has a well-defined amount of energy, capable of holding it in a certain orbit which, therefore, is called the energy level of the orbit. Consequently, the orbit represents the path that the electron travels around the nucleus while the energy level represents the amount of energy sufficient to keep it in that orbit.

As we have seen, quantum mechanics is able to describe the behavior of very small particles, such as electrons, and we have also seen that the rules of classical physics do not apply to these particles. In the quantum vision a new, more complex atomic model has been introduced, in which we no longer speak of orbit but of orbital.

Orbit and orbital.

Quantum mechanics has amply demonstrated that it is not possible to define the trajectory of an electron, so the term "orbit" is not appropriate. It is necessary to abandon the concept of an orbit traveled by an electron to replace it with that of "orbital", also defined as a "probability cloud". This happens because we cannot establish with certainty what is the position of the electron within the atom at a given instant or in the next instant. Through complicated mathematical functions it is possible to obtain information on the position of the electron, but only in terms of the probability of its presence in a certain space around the nucleus (Fig. 41).

In other words, while the orbit is a defined trajectory, the orbital is a mathematical function that allows you to establish where the electron is most likely to find in the space around the atom.

All this is a direct consequence of a fundamental principle of quantum mechanics discovered by the German physicist Werner Heisenberg the Uncertainty Principle or Heisenberg Principle, according to which it is not possible to simultaneously know the position and speed of the electron in its orbit. This confirms the peculiarity of the behavior of elementary particles compared to that of bodies at the macroscopic level, that is, at the level of our dimensions and our experience.

At our level, in fact, it is possible to calculate with extreme precision the trajectory and speed of a stone even before launching it, just as it is possible to calculate the path of celestial bodies or the travel itinerary of space satellites. At the subatomic level, however, this ability becomes impossible. Heisenberg's principle once again establishes the mystery-cloaked diversities of particle behavior.

As regards not only the electron but any other elementary particle, we can only define a certain region of space where the particle could be found; it is this region of space that is called orbital, and is determined by the energy possessed by the particle.

We can imagine the orbital as a compromise between the repulsion of the various electrons between them and the attraction exerted by the nucleus. Therefore, the more protons and electrons an atom has, the more complicated the situation of the internal relations of attraction and repulsion becomes; the atom becomes a crawl space in which wasps move at unimaginable speeds along probabilistic orbitals.

Under these conditions, the only way to know very roughly where an electron is and at what speed it moves is to treat it as a wave; the electron will be in any point X of the wave that describes it. This principle is defined by a mathematical function, the wave function, discovered by the Austrian physicist and mathematician Erwin Schrödinger.

This function, in fact, does not define the precise coordinates of the electron, but delimits a volume of space within which it can be found with greater probability.

Today we know that quantum laws can be applied not only to the electron, but also to the particles present in the nuclei of atoms, protons and neutrons.

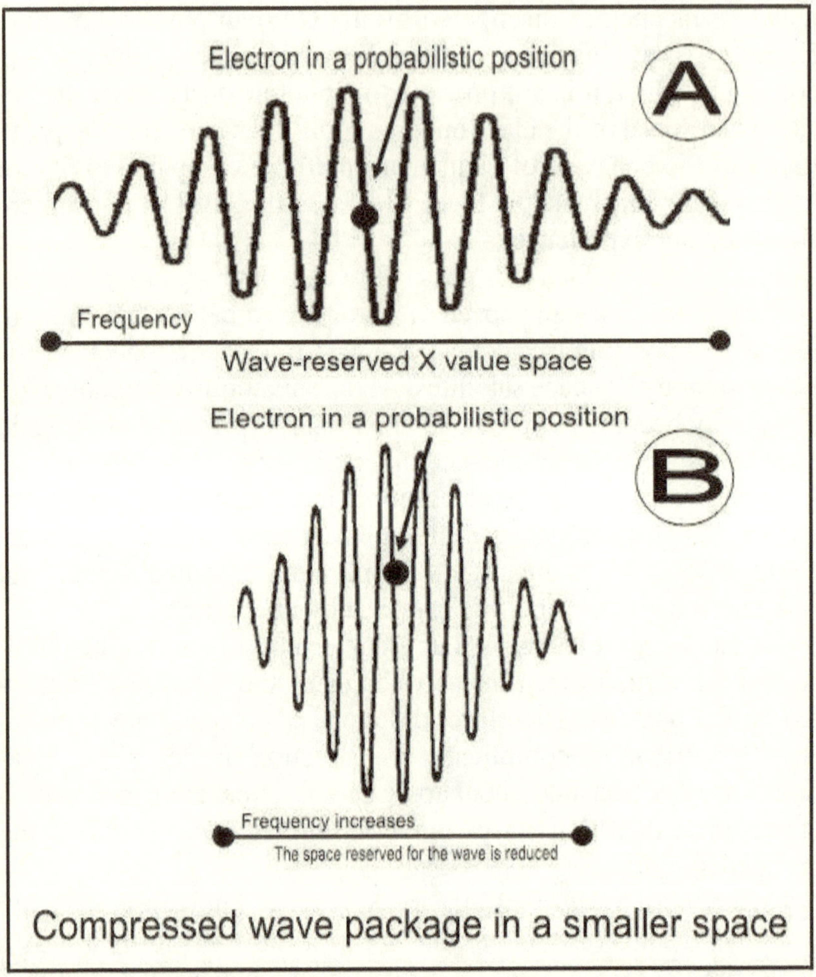

Figure 41 - Compressing the space region of the wave increases its frequency and therefore the rotation speed of the particle.

A In the space of value "X" the frequency of the wave has value "a".

B In a narrower space the wave frequency has a value "b", higher than "a"

Dynamic view of the atom.

What has been said so far provides a much more dynamic view of the atom than the classical one. In fact, considering an electron as a particle, we cannot know where it is exactly, except that it is placed in any point of the wave function that describes it (fig. 41A).

Modern physics has come to conceive the universe as an inextricable network, within which relationships are absolutely dynamic, not static. In quantum mechanics this dynamic aspect, if possible, is enhanced and is a direct consequence of the wave nature of the particles. Consequently, the properties of subatomic particles can only be understood in the context of their dynamism.

The fact that the particles are also waves causes apparently bizarre behavior that must be taken into account. If we force a particle into a certain space, it reacts by agitating in that space. The agitation is much faster the more the space is restricted.

This property becomes understandable by examining Figure 41.

In the upper part the wave function of an electron is depicted, enclosed in a region of space of X value with the electron placed in a probabilistic position.

In the lower part of Figure 41B, by restricting the available space, the wave vibrates at a higher frequency and this causes an increase in the speed of the electron. The smaller the confinement region, the more the electron's motion will increase.

The tendency of particles to react to each confinement by increasing the speed generates the state of agitation of matter, typical of the subatomic world.

Moreover, even in the nucleus there is no static; indeed, if possible, the agitation takes on paroxysmal forms. We can understand this if we consider that the volume of the nucleus is a thousand billion times smaller than the total volume of the atom. In fact, as we have already mentioned, if the nucleus were 3-4 millimeters in size, the outermost electron would rotate in a spherical volume with a diameter of about 100 meters. In the volume of space of such an incredibly small nucleus two forces come into play: on the one hand, the repulsion between the protons, which all have a positive electric charge, tends to "explode" the nucleus into space. On the other hand, however, the strong nuclear interaction that exerts attraction between all particles composed of quarks comes into play. Therefore, the core vibrates, in an eternal conflict made of repulsion and attraction, with a speed exponentially enhanced by its very small size.

"Things" or events?

Since particles are part of the larger structures such as molecules and larger aggregates of matter, these too have a tendency to move in a perennial dance, that is, they are naturally agitated. Quantum mechanics demonstrates how matter is never inert, but constantly in motion. Although on a macroscopic level the objects appear immobile and solid to us, in their intimate composition they seethe with frenetic activity.

The objects of our daily reality are composed of atoms that connect to each other to form all the molecular structures we know. These vibrate in harmony with the ambient temperature. The electrons are bound to the atoms by electric forces of attraction and repulsion which squeeze them towards the nucleus thereby causing a very high rotation speed. In the same nucleus, protons and neutrons are restricted in such tiny volumes as to cause them to agitate at speeds not even imaginable.

The further one goes into the study of the atom, the more it becomes evident that it is not the ultimate component of matter, solid and compact, as suggested by Newton, but a set of activities, i.e. vibrations due to particles that are simultaneously waves, in eternal frenetic movement in a predominantly empty space.

Karl Raimund Popper, Austrian philosopher and scholar of the methods by which to achieve certain and scientific knowledge of things, attributes to metaphysics, if correctly understood, the function of stimulating progress. Physical determinism is harshly contested by Popper, both from the scientific point of view and as a presupposition of totalitarianism. In his book "The I and his brain" Popper expresses his thoughts on the material universe in these terms:

> "Matter turns out to be highly concentrated energy, which can be transformed into other forms of energy and therefore something of a processual nature, since it can be converted into other processes such as light and, obviously, movement and heat. We can therefore say that the results of modern physics suggest the need to abandon the idea of a substance or essence.
>
> Now (as A. N. Whitehead pointed out above all) the universe appears not as a collection of things, but as a set of interacting events and processes.
>
> Therefore, a modern physicist may very well say that physical things - bodies, matter - have an atomic structure. But in turn the atoms have a structure that can hardly be defined as "material" and certainly not "substantial": in pursuing the programmatic intervention to explain the structure of matter, physics had to transcend materialism. All this development beyond materialism arose from the research conducted on the structure of matter, within atoms, and was therefore a result of the same materialistic research program. This development has left intact the importance and reality of matter and material things: atoms, molecules and structures of molecules. Indeed, it could be said that their reality has gained from it, since, as the history of materialism and especially of atomism demonstrates, the reality of matter was considered doubtful not only by idealist philosophers such as Berkeley and Hume, but also by physicists such as Mach , precisely at the time when quantum theory was emerging."

In his theory of quantum electrodynamics (defined as the jewel of physics), physicist Richard Feynman suggests that from the Heisenberg uncertainty principle it derives that the quantum vacuum, also called the zero-point field, is permeated by a sea of fluctuations that create pairs of virtual particles and antiparticles, which annihilate in a time inversely proportional to their energy.

Einstein's famous formula E = mc2 establishes that matter and energy are convertible to each other. Therefore, energy can take many forms: it can be classically mechanical in two modes, kinetic or potential; moreover, it can be chemical, electromagnetic, gravitational, thermal, nuclear. All these expressions of energy are components of the same global energy, which can change shape but cannot circumvent the rule that the quantity always remains the same. In its immutability and convertibility, energy plays the role of connecting all physical phenomena in a network of transformations, but any physical transformation must necessarily provide that the amount of energy, before and after, remains the same. Therefore, energy cannot be destroyed but can take on different forms; indeed, it is in constant transformation.

Locality and non-locality.

The locality defines the area in which there are manifestations of energies and events delimited by the laws of classical physics; in the local reality there is causality or determinism (each event is determined by a previous event), all reality is subject to the arrow of time that moves only forward and, in any case, it is not possible to exceed the speed of light so there can be no instant communication of information over long distances. Moreover, any energy that carries any information, undergoes a decrease in power that is much higher, the further it moves away from the source, as happens for a light or a sound that becomes increasingly dim as the distance increases.

In the non-local domain, on the other hand, there are events and manifestations that do not undergo the typical restrictions of the local domain: they are not limited either by time or by distance. In the non-local level it does not exist yesterday and today, before and after, but only "now and always". Likewise, there is no high and low, near and far, but only "here and everywhere".

Entanglement.

Entities involved in an "entanglement," particles or photons, are linked together because they were produced by a process that bound them together in a special way. For example, two photons emitted by the same atom. Such photons or particles will remain "entangled", ie coupled, forever. If in some way one acts on one of the two, his "twin", wherever he is in the universe, will react instantly in turn
(Amir Dan Aczel, Israeli mathematician and physicist)

The wave behavior of matter is the basis of another surprising quantum phenomenon known as entanglement, an English term that can be translated as intertwining. Entanglement represents one of the most mysterious, and still substantially unexplained phenomena in all of modern physics. According to Erwin Schrödinger, one of the founding fathers of quantum mechanics, it is undoubtedly the "characterizing feature" of the whole theory.

Entanglement has such incredible properties that they caused Einstein to doubt the reality of the phenomenon. In 1935 together with Boris Podolsky and Nathan Rosen he formulated the famous "EPR paradox" (so called from the initials of the surnames of the three scientists) which, in the very meaning of the name, highlighted the phenomenon of entanglement as paradoxical.

Indeed, the EPR paradox, first considered as a valid refutation of entanglement, contained a fundamental error, highlighted in 1964 by John Stewart Bell and confirmed in 1982 by Alain Aspect's experiments. Subsequently, in 1998, Anton Zeilinger and others improved these experiments and confirmed their results.

Today we can conclude that in countless studies quantum physics has been shown to be able to produce correct experimental predictions up to an unprecedented precision, and the correlation phenomena between particles, typical of quantum entanglement, have actually been observed. Currently this theory is the basis of emerging technologies such as quantum computers and quantum cryptography and has made it possible to carry out experiments related to teleportation, on which the expectations of the technological world are concentrated.

The superposition principle.

In the examination of the double slit experiment conducted by sending only one particle at a time, we saw how this crosses both slits and creates an interference phenomenon. It is evident that the particle, being the only one present, interferes with itself. This is an astounding but real phenomenon, which takes the name of the "quantum principle of superposition of states"

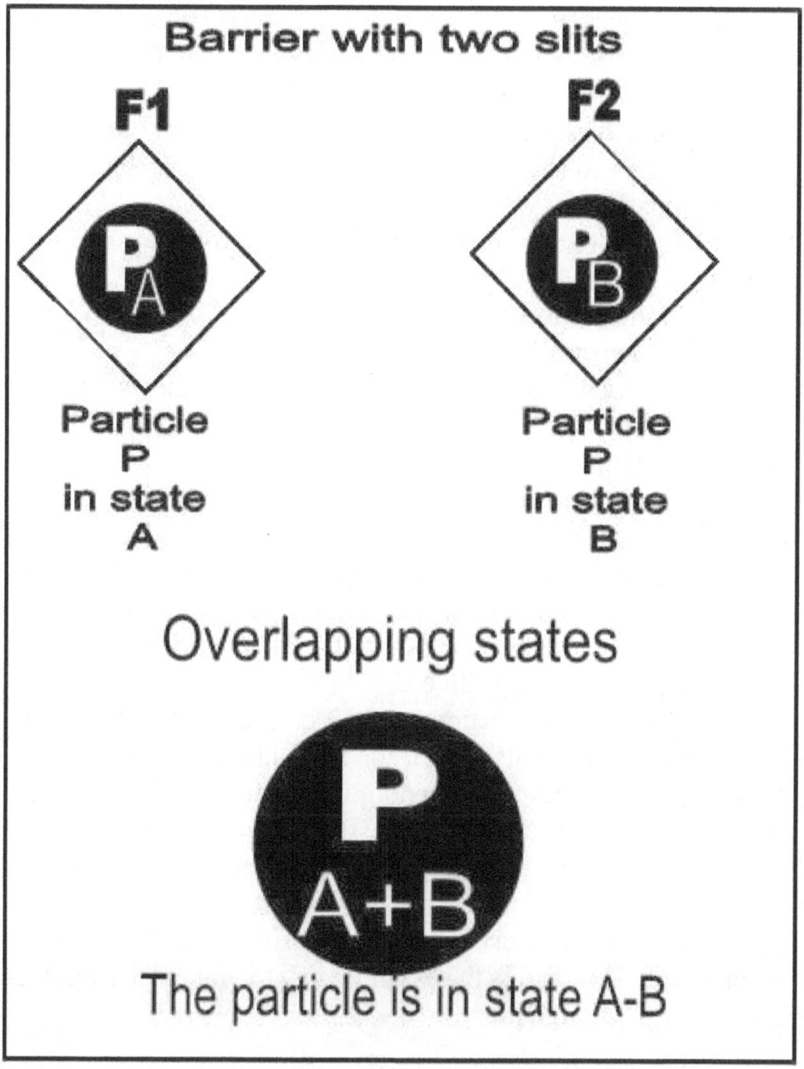

Figure 42 - The P particle passing through both slits is in a quantum superposition state.

F1 The P particle passes through the F1 slit and has the state Pa

F2 The same particle P also passes into the slit F2 and has the Pb state.

The particle P is in "superposition of states". Its state is P (a + b)

The superposition principle states that, just like the waves of classical physics, two or more quantum states can be "superimposed", ie added together, and the result will be another valid quantum state.

Therefore, the state of a system can be composed of two or more states, in order to share some of the properties of each of the states that compose it.

If we mark with A and B two different properties of the same particle, such as:

state A: the particle is in the place F1

state B: the particle is in place F2

then the superposition of the states of that particle can be written as A + B because it has something in common with both state A and state B (fig. 42).

The particle is in state A when it passes through the slit F1, and in state B when it passes through the slit F2. Since the particle passes through both slits simultaneously, it is simultaneously in the two states, that is, in a superposition state, which is written A + B and is a combination of the two states "The particle is in state A" and "The particle it is in state B ".

If we observed the particle while it crosses the barrier we would see it pass through only one of the two slits and we would not obtain the interference phenomenon; if, on the other hand, we do not observe it, it passes through both slits, generates the state of superposition and signals it by producing the interference figures on the screen behind it.

The superposition principle is one of the greatest mysteries of quantum physics, and is the essential key to understanding the phenomenon of entanglement.

The correlation.

The correlation can be considered the result of entanglement: two particles that have been made entangled with each other are said to be related. This means that the particles are connected to form an intertwined system, and that they react in unison, each to the events that affect the other. They behave just like twins would do, for example those whose stories I presented at the beginning of this book.

Thus, two related particles can be considered twins. Just as human twins are involved in the same birth, elementary particles become entangled when they are involved in a phenomenon that interests them in a non-trivial way.

The cases of the McWhirter brothers and the Furness sisters can be considered extraordinary because the protagonists reacted to the events that affected their twin without any apparent cause, thus upsetting the principle of causality according to which any event that occurs in the universe must be the cause of another. event that occurred previously. In fact, even the related particles react each to the events that affect the other without any causal link. In fact, the reaction of one of the twins occurs when this is totally unaware of what is happening to the other.

Even in the case of entangled particles the same thing happens: the changes in state are not one after the other, but simultaneous; and this, mind you, at whatever distance the particles are.

This circumstance constitutes a real puzzle for physicists since it can mean only two things: either the particles communicate with each other by transmitting signals at a practically infinite speed, and in any case higher than that of light, or there is another form of communication that is independent of the space and time and is therefore purely psychic, of the same form of thought, capable of instantly wandering anywhere in the universe, from the past to the future, without any physical limit.

However, even this example is imperfect because it would foresee that one of the two particles would launch "a call" to the other, in any case, albeit at the speed of thought.

Instead, the two particles decide their variation at the same time; neither of the two particles decide this first. If that happened, causality would come into play - albeit at extreme speed. Not so: particles behave just as if they were a single particle.

Entanglement is a bond of coherence.

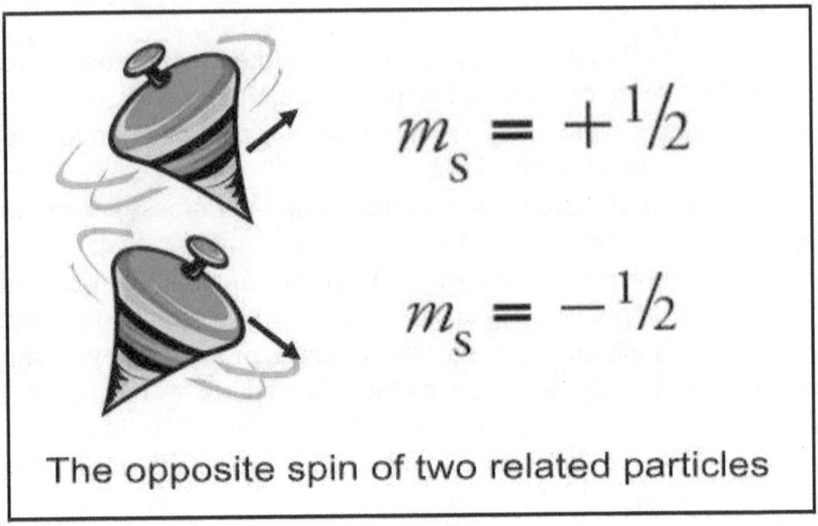

Figure 43 - Spin (rotation) was initially introduced to indicate the moment of the electron's intrinsic momentum.

For example, if two photons with opposite spin are correlated forming an entangled system, the variation of the spin (fig. 43) of one affects the spin of the other at the same time. There is no way to verify which variation occurred first.

Figure 31 illustrates a method used in laboratories to obtain correlated particles. At point A, an electron, placed on any orbit, has an energy of 10 (an example value). At point B the electron jumped to an inner orbit; now the energy required is lower, and we can consider it equal to 9. The unnecessary quantity, (a quantum), served to generate a photon.

To obtain two correlated photons, an electron will double jump.

Let's imagine that we have obtained two correlated photons in this way and examine the relationships between them.

Superposition of two particles.

We assume that each of the two particles (P1 and P2) can be in two places at the same time and therefore has two states. The particle P1 will possess the states P1A and P1B, while the particle P2 will possess the states P2Y and P2Z. Recall that states A, B, Y and Z represent different places where particles can be (Figure 44).

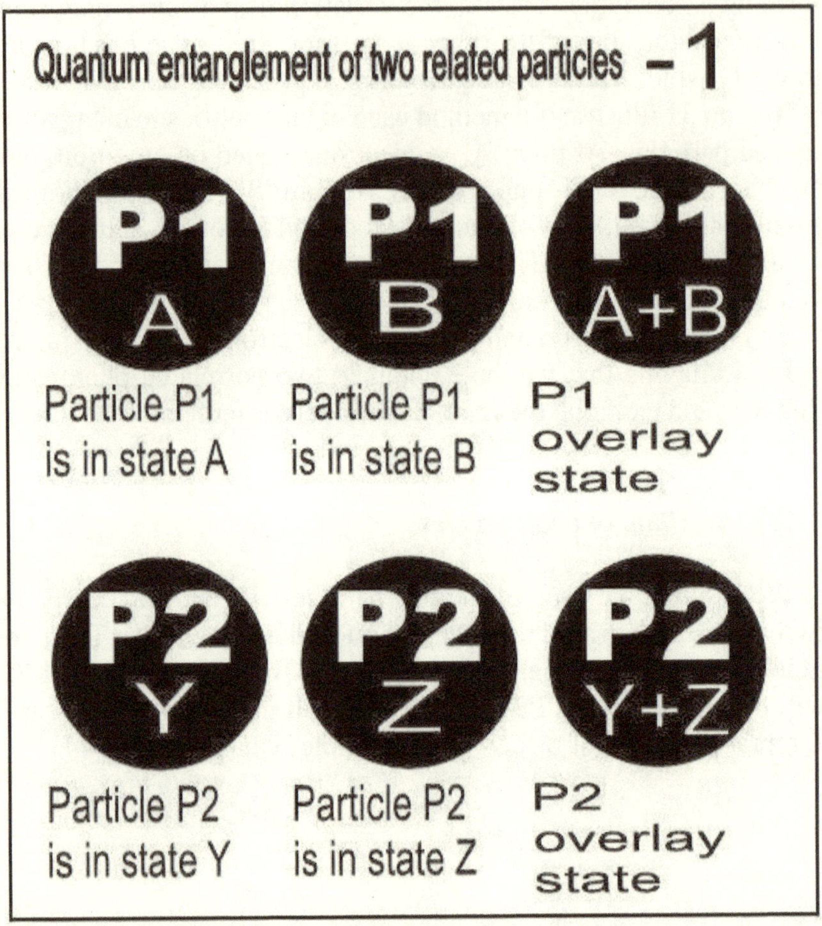

Figure 44 - Quantum entanglement of two correlated particles. (First stage)

The two particles P1 and P2 both possess two states.

The particle P1 is superimposed on (P1 (a + b)

Particle P2 is superimposed (P2 (a + b).

The two related particles have the property of possessing two states. Now imagine that the particle P1 is in the state A, and the particle P2 is in the state Y. The result is a system of two particles with the state AY, which is called the "produced state". If, on the other hand, we imagine that the particle P1 is in state B and the particle P2 is in state Z, the result is the product state BZ (Figure 45).

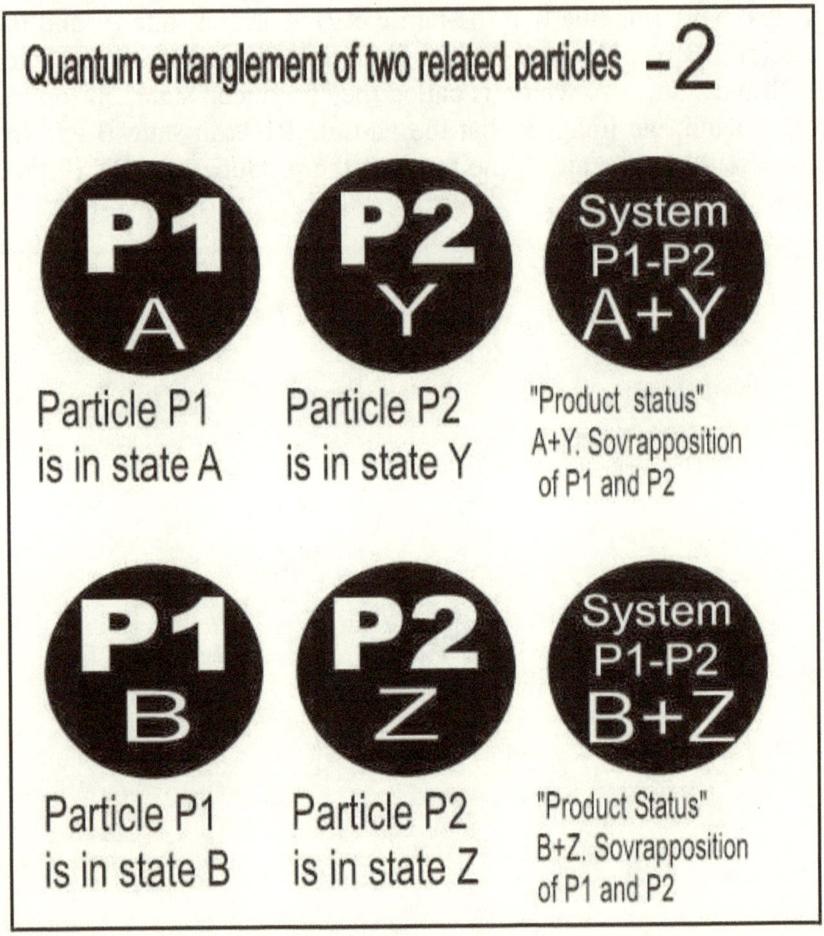

Figure 45 - Quantum entanglement of two correlated particles. (Second phase)

"Product state" of related particles P1 and P2

The two particles P1a and P2y produce the superposition state P1-P2 (a + y)

The two particles P1b and P2z produce the superposition state P1-P2 (b + z)

.

At this point, with a further step, we can apply the superposition principle to the set of the two systems AY + BZ.

This state is called the "entangled state" (Figure 46).

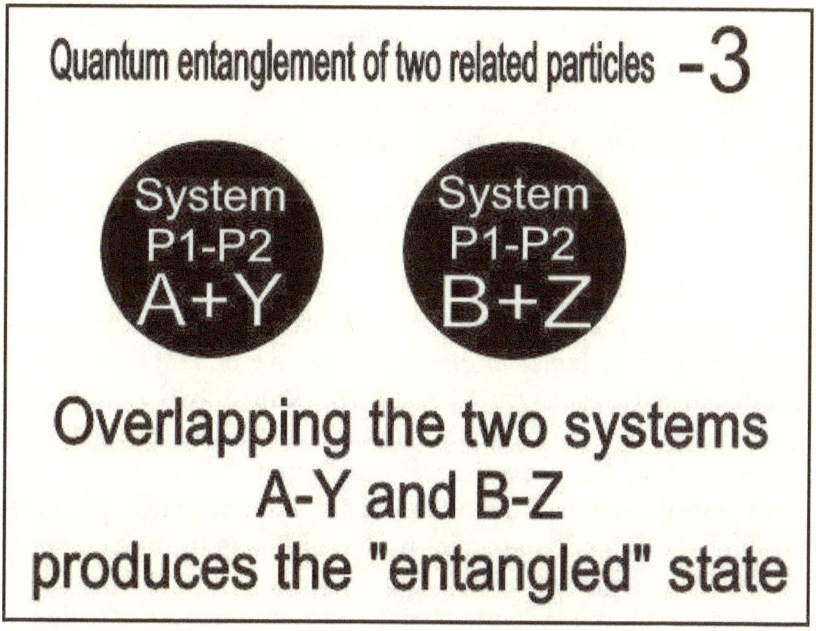

At this point, with a further step, we can apply the superposition principle to the set of the two systems AY + BZ.

This state is called the "entangled state" (Figure 46.

The overlap of the two systems is called entanglement. The product state AY indicates that the P1 particle is in the A state, and the P2 particle in the Y state.

Similarly the product state BZ indicates that the particle P1 is in the B state and the P2 particle in the Z state.

The entangled state AY + BZ, which represents a superposition of states, indicates that there are mandatory conclusions regarding the two particles.

In fact, if you make a measurement which shows that the particle P1 is in the state A, then the particle P2 must necessarily be in the state Y, and vice versa.

Similarly, if, by carrying out the measurement, it is found that the particle P1 is in the state B, then the particle P2 must necessarily be in the state Z.

In other words, when two particles are in the entangled state, it is not possible to consider one apart from the other. Ultimately, the entangled state results from the superposition of two product states.

Once two particles are correlated, a state of entanglement is produced between them which takes no account of time or space. Each intervention on one of the two particles simultaneously produces a reciprocal change in the other, wherever it is in the universe.

The Danish physicist Niels Bohr, founder of the famous Copenhagen School, made fundamental contributions to the understanding of the structure of the atom and quantum mechanics, and for these he received the Nobel Prize in Physics in 1922. Regarding the correlation of particles, he said:

> "Even if two related photons were on two different galaxies, they would always continue to remain a single entity, and the action performed by one of them would also affect the other".

With this, entanglement solemnly proclaims that the universe is not all subject to the laws of classical physics, but, at least at the subatomic level, these laws are no longer valid.

The fact that the two entangled particles continue to behave in unison whatever the distance between them poses an apparently unsolvable problem.

How do the two particles communicate with each other? Certainly not with signals sent with classical methods; these, in fact, would fade with increasing distance and in any case could not exceed the speed of light. There must certainly be a possibility of non-physical communication.

Non-locality and extrasensory phenomena.

The most surprising and intriguing feature that emerges from what has been said so far is that the subatomic particles are able to exchange information instantaneously, that is, they are connected in a non-local way; Put simply, the information does not cross any physical place to connect the two particles; the non-local level is a purely psychic level (to avoid using the term spiritual), that is, made up of pure information, or intelligence.

It could also be argued that the two particles are separated only in appearance remaining, in fact, one thing, since they behave just like that. This does not solve the problem. If the two particles were still one thing then the whole universe should be one, that is, it would not be made up of separate objects (if not at the macrophysical level) while at the subatomic level it would be a single reality, that Unus Mundus that it emerged from the collaboration of Jung and Pauli or the implicate universe theorized by David Bohm. Even in this case, however, a single universal intelligence would be required to govern the Cosmos.

Albert Einstein strongly opposed the principle of non-locality because it violated the physical laws that he himself had formulated, all based on locality. He constantly tried in his life to develop a theory capable of unifying the concepts of classical physics with those of quantum; unfortunately, he did not succeed.

In 1982, following the experiments conducted by Alain Aspect, the theory of entanglement and with this the concept of non-locality was definitively confirmed. Since then we have known that the "local realism" that we perceive in our dimension is not an absolute law of nature.

At this point it becomes possible to imagine that space and time are inconsistent and exist only on a perceptual level. Aspect's experiment showed us that there is a connection between particles that goes beyond space-time. Moreover, since the whole universe is made up of particles, we can ask ourselves whether the phenomenon of non-locality should be confined only to subatomic reality or whether it can also occur between larger objects, or even between living beings.

Is it possible for biological organisms, such as humans, to access the non-local level? If the answer were yes, then we could imagine an exchange of information between different organisms by unconventional means, through that mysterious dimension that is non-locality.

In fact, this possibility should not be excluded, on the contrary it exists and is very widespread and well known. It is called telepathy.

A characteristic of telepathy is that it works best between people who are linked by particular bonds of affection or kinship, such as mothers and children, that is, "entangled" with each other emotionally.

Our brain is certainly capable of elaborating wishes to connect with the people closest to us. Perhaps non-locality represents the hitherto unknown means by which such communications can actually take place. Even if human beings do not know how to use telepathy fully, they still have the means to do so, that is, a mind capable of processing thoughts and desiring their transmission. Sometimes this happens spontaneously, that is, it happens even if we can't explain how it happens.

Telepathy probably consists of a phenomenon of correlation between minds. Consciousness, in the exercise of its ability to develop intentions, can correlate two brains putting them in an entangled state.

Another aspect that could be linked to non-locality is that relating to the phenomena of foresight and premonition.

Indeed, in the non-local level one could recognize great correspondences with the collective unconscious, the cosmic level theorized by Jung containing all the history and wisdom of humanity in its evolution. The possibility of an "entanglement" that develops on a non-local level coincides with the possibility of a collective level of consciousness and makes it plausible that individual consciences can access its past, present and future contents.

In this sense, the experiments of Alain Aspect and his collaborators could also be considered experiences of quantum metaphysics. By demonstrating non-locality, they also demonstrated the existence of a level transcending the reality of space-time, and disavowed the hypothesis of a solely material universe.

Brian Josephson is a Welsh physicist, born in Cardiff in 1940. When he was only 22 he discovered the Josephson effect, and as a result of this he received the Nobel Prize in Physics in 1973.

Josephson is one of the most famous scientists among those who support the possibility that parapsychological phenomena are real. He particularly ventured into the study of Eastern mysticism, and the possible implications that it could present in relation to the interpretation of scientific phenomena in a mystical or metaphysical key.

In particular, Josephson states that one of his guiding methods is: "nullius in verba" (Don't trust anyone's words) with reference to the fact that:

> "... if all scientists reject an idea, this should not necessarily be considered as proof that the aforementioned idea is absurd; on the contrary, it is necessary to carefully examine the alleged motivations of such opinions and judge how well they hold up to a careful examination."

An "entangled" universe.

Brian Greene is an American physicist, one of the most important scholars of string theory. In 2014, his book "The plot of the cosmos. Space, time, reality" was published in the Italian version.

In this book, focusing on the riddle of time, Greene establishes that no physical rules confirm the idea that time flows in a particular direction. Furthermore, among the themes of this book, in addition to a basic exposition of string theory, you can find the discussion of quantum entanglement. It is essentially an analysis of the nature of reality with excursions on topics such as spacetime, cosmology and the origins of the universe (the so-called theory of everything).

Greene believes, in fact, that the universe is a whole intrinsically entangled and consistent with itself.

Another authoritative confirmation comes from the book "The Non-Local Universe: The New Physics and Matters of the Mind", written by two physicists, Robert Nadeau and Menas Kafatos. Here is a quote often taken from this text:

> "All the particles in the history of the cosmos have interacted with other particles in the manner detected by Aspect's experiments. Virtually everything in our immediate physical environment is made up of quanta who have interacted in this manner with other quanta from the big bang to the present moment. (...) It must also be considered that quantum entanglement grows exponentially with the number of particles involved in the original quantum state and that there is no theoretical limit to the number of these entangled particles. If this is the case, the universe at a very basic level could be a vast network of particles, which remain in contact with each other over all distances and in zero time and in the absence of energy or information transfer. This suggests, however strange or bizarre it may seem, that

> all physical reality is a Single Quantum System that chorally responds to further interactions. "

In fact, if we consider the Big Bang theory that I described above, its basic assumption turns out to be the very powerful explosion, which occurred about 15 billion years ago, of a point-like singularity of energy from which it would have originated, thanks to an expansion still going on, the whole universe. This means that the entire universe was initially compressed into a reality of almost infinite density. The fact that all the particles were so closely connected and that they were all born from the same explosive event, makes it legitimate to think that all the particles of the universe, in reality, are entangled with each other.

Not only that, it is possible to hypothesize that all the particles of the universe retain memory of their twinning with every other particle. This means that the separation of reality into "things" is only an illusion. The separation is due to the fact that we perceive the macroscopic level of the universe. In reality, the universe, although expanded into billions of galaxies distant from each other billions of years, would still be united in itself with non-local bonds, beyond any space and any time. If this could be demonstrated it would also be demonstrated that we all live in an immense holistic reality where every single part is worth the whole because it includes the whole.

Certainly, there are still many things that we do not know and that can even surpass our imagination.

Confirmations of entanglement.

Einstein was wrong when he said, "God doesn't play dice." In fact, the consideration of black holes suggests not only that God plays dice, but that he sometimes confuses us by throwing them where they cannot be seen.
(Stephen Hawking)

Someone will probably remember "Noein". It is a 2005 sci-fi animated series produced by Satelight studio, consisting of 24 episodes, conceived and directed by Kazuki Akane.

Noein, the character from which the series takes its name, is an evanescent figure, characterized by a golden mask with one eye. He is the one who pulls the strings of events from the very beginning. He is deeply imbued with theories of quantum mechanics.

Noein treats the concept of time as a dimension that he resonates with other timelines. In particular, he sees the universe branching out into an infinite number of probabilities, favoring Hugh Everett's interpretation of the many worlds. However, he also incorporates many aspects of Bohr's and the Copenhagen School's interpretation that the presence of an observer is essential to determine the collapse of the probability wave in a measurable situation.

Haruka, a female character, has the power of an observer within the multiverse, a power that enables her to determine the outcome of an action simply by "observing" it.

In the episode "Overlaps," Uchida explains Schrödinger's cat paradox to her bodyguard Koriyama. This is an ideal situation where only an observer is able to determine the cat's death the moment he opens the box. In the same episode, reference is also made to the phrase "God does not play dice", a famous quote from Albert Einstein.

> "God doesn't play dice, did you know that? Let's say you throw a die, and from that roll comes the number six. "
>
> "I'd prefer number one, because it's the first."
>
> "Okay, one or six is the same. So, let's say number one is out. The number five could also come out, or the number four could come out. They have the same odds, don't they?
>
> "Indeed, yes."
>
> "Can you imagine a world where number five came out and another world where number four came out? Is it possible that there are countless worlds in which

countless copies of yourself throw an infinite number of dice?

"God doesn't play dice"

This quote is one of Einstein's most famous maxims, who repeated it several times on various occasions. The scientist used it in opposition to the nondeterministic implications of quantum mechanics, advocated by Niels Bohr and his Copenhagen School.

The fact that the position and velocity of an electron could not be established with certainty, and that these data could only be expressed in terms of probability, greatly disturbed Einstein's rigorously scientific sense, firmly based on the certainties of locality and determinism.

According to Bohr's theory, throwing an electron several times, under the same experimental conditions, could not be expected to always give the same result, because its trajectory would always be subject to the unstable cloud of probability. A bit as if, throwing a die with the same mechanism, one always had to expect the same number to come out; but we know that this does not happen, due to the countless variations that affect the launch, some very unstable such as the temperature and humidity of the environment. With his aphorism Einstein meant that quantum mechanics could not be considered similar to the roll of the dice, but it must necessarily provide reliable results in his calculations.

There are many quotes in which Einstein strongly expresses his doubts. On December 4, 1926 he wrote these words in his diary:

> "Quantum mechanics is worthy of all respect, but an inner voice tells me that it is not yet the right solution. It is a theory that tells us many things, but it does not let us delve deeper into the secret of the Great Old Man. In any case , I am convinced that the Great Old Man does not play dice with the world. "

In 1942, writing to Cornelius Lanczos, a Hungarian mathematician and physicist and a great friend of him, Einstein said:

> "How difficult it is to get a glimpse of God's cards. But I don't believe for a single moment that He plays dice or uses" telepathic "methods as current quantum theory would like."

In particular, the numerous exchanges of views between Einstein and Niels Bohr, supporter of the opposite thesis, are known. Many witnesses and Bohr himself recall the verbal confrontation that took place at the Solvay Congress in 1927. Bohr writes:

> "For his part Einstein asked us, jokingly, if we could actually believe that divine Providence had" played dice ". I replied that the ancient thinkers had warned us to be very cautious in defining the attributes of Providence in common language."

Werner Heisenberg was also present on that occasion and remembers Einsteinm's joke to which Bohr replied: "But it's not up to us to tell God how he must make the world go!".

In 1944, on September 7, Einstein wrote to Bohr as follows:

> "Our scientific perspectives are now the polar opposite. You believe that God plays dice with the world; I believe instead that everything obeys a law. I believe in a world of objective realities and I try to grasp them in a furiously speculative way. I hope that someone will discover a more realistic path than mine. Even the great initial success of quantum theory fails to convince me that randomness is at the root of it all. However, I am

> well aware that younger colleagues consider this attitude to be an effect of sclerosis . One day we will know which of these two instinctive attitudes was the right one. "

Einstein's thesis was based on the concept of hidden variables. According to him, if the Copenhagen School version of quantum mechanics had yielded nondeterministic results, it certainly depended on the fact that some important parameter had been overlooked

The EPR paradox.

Quantum entanglement, deriving from the Copenhagen interpretation, was considered paradoxical because it was considered incompatible with special relativity which considers the speed of light as the maximum at which any type of information can travel. Furthermore, quantum entanglement was incompatible with the locality principle. From this arose Einstein's belief that quantum theory was incomplete. According to him, the theory included hidden variables, that is, parameters that were not taken into account in the calculations.

In 1935, together with Boris Podolsky and Nathan Rosen, Einstein proposed a thought experiment in the essay "Can Quantum-Mechanical Description of Physical Reality Bẹ Considered Complete?" (Can the quantum description of physical reality be considered complete?). The essay was published in the "Physical Review" XLVII, 1935, pp. 777-780).

In this work, known as "Paradox EPR" from the initials of their names, the three creators intended to demonstrate that quantum mechanics led to valid results, but produced paradoxical consequences. Consequently, according to them, this was not a complete physical theory.

Five months later, Niels Bohr responded by publishing in the same journal, an article similarly titled (Physical Review XLVIII, 1935, p. 700).

In reality, at the time the clash between the two scientists had no winners, above all because the tools of the time did not allow a real experimental verification of the arguments supported. As is still used today, the proposed "experiments" were called "mental experiments" or "Gedankenexperiment", according to the German term coined by the Danish physicist Hans Christian Ørsted. Therefore, a thought experiment is still aimed, even today, at testing, corroborating or refuting a hypothesis or theory, but it has an imaginary and not empirical character because the practical limitations make it impossible.

It should also be said that the EPR experiment had a degree of complexity such as to make the conclusions that could be drawn from it doubtful in any case.

Despite this, Einstein was not the only one to have doubts, but other scientists in their work also highlighted the paradoxical sides of the Copenhagen interpretation of quantum mechanics.

chrödinger's cat paradox.

In 1935 Erwin Schrödinger, an Austrian physicist who had received the Nobel Prize in 1933, published the article in which he described the famous "cat paradox", trying to show how bizarre the superposition theory was. Indeed Schrödinger by proposing this paradox intended, like Einstein, to demonstrate the limits of quantum physics, although he himself had been one of the founders of this science. As already said, according to quantum physics, the behavior of an elementary particle cannot be predicted exactly, but only in a probabilistic way.

This is not the case for macroscopic systems, that is, made up of millions of aggregate atoms: once certain data is known, it is possible to calculate exactly what will happen in them, and when it will happen.

Schrödinger's (mental) experiment involves a cat being closed in a box containing a radioactive substance. We know for sure that a single atom of such a substance will emit radiation, but we cannot absolutely say when this will happen.

Figure 47 - Above, Erwin Schrödinger in a vintage photo and an Austrian commemorative stamp. Below, a joking post on the famous cat paradox, from the blog "Cenacolo Jung-Pauli

In the box, together with the cat, there is a device whereby, when the radioactive substance emits a radiation, this will cause the rupture of a cyanide vial which will kill the cat

Certainly times have changed and a scientist who proposed such an experiment today would immediately find the laboratory besieged by animal rights activists. However, we know that Schrödinger never put any cat in any box, indeed, probably, he owned a cat he was very fond of.

Returning to the experiment, we know that the death of the animal will occur only if the radioactive substance emits a radiation, but we do not know whether this will happen or when, eventually, it will happen.

The box will be kept closed for one hour. Knowing that there is a 50% chance that the atom will decay during this hour, we can establish that we have the same chance of finding the cat alive or dead when we open the box.

According to quantum mechanics in the Copenhagen interpretation, if the behavior of the atom is not explicitly observed, the latter is in a superposition of states, that is, it has decayed but, at the same time, it has not decayed; the atom remains in this superposition until someone opens the box causing the "collapse of the wave function".

Applying Schrödinger's paradox to quantum mechanics, as long as the box remains closed we have an atom both decayed and not decayed, a vial of cyanide both whole and broken and a cat both alive and dead.

On the other hand, when the box is opened, the system collapses into a defined state. This is similar to what was said in the case of the double slit experiment: when a photon is launched it behaves like a wave, passes through both slits and causes an interference phenomenon. If, on the other hand, we apply an instrument to one of the two slits to observe its passage, the wave function collapses and the photon behaves like a particle, passing through a single slit.

Therefore, only when the box is opened does the system composed of radioactive atom, cyanide vial and cat assume a non-superimposed, unique and coherent state.

Jokingly we can say that from this situation the awareness would arise that, by opening the box, it would be probable to determine the death of the cat. Instead, the cat would continue to be alive (or at least both dead and alive) as long as the box was closed.

Obviously, Schrödinger's cat paradox describes a situation that is absolutely far from reality so that, in defense of quantum physics, we can say that in a certain sense Schrödinger cheated by putting together two realities, the quantum and the macroscopic. Of course, we have not yet reached the unification of the different physical laws of the two levels.

John Stewart Bell's inequality.

Erwin Schrödinger was not alone in supporting Einstein's doubts. Many other scientists were conceptually critical of the Copenhagen School thesis. However, much of the scientific community had gradually become convinced that Einstein's position was wrong. But the argument had been shelved and, albeit without evidence, the majority of scientists recognized the validity of Bohr's "non-local" theory.

A revival of the debate took place thanks to John Stewart Bell, (fig. 48) a quiet, freckled, red-haired young man born in Belfast, Northern Ireland, in 1928.

Bell had graduated in experimental physics from Queen's University in Belfast in 1948. After several years, spent in various institutes, he had started working at CERN, (European Organization for Nuclear Research), the largest laboratory in the world for particle physics, located on the border between Switzerland and France, on the western outskirts of the city of Geneva in the municipality of Meyrin.

Over the years he spent at CERN, Bell developed his research in theoretical physics and worked in the field of particle accelerators. Currently the CERN complex includes 7 main accelerators, built in various periods starting from the foundation of the institute.

Figure 48 - John Stewart Bell proposed an experiment, carried out in 1982 by Alain Aspect, which demonstrated the inaccuracy of the theory expressed by Einstein together with Podolski and Rosen (EPR paradox), confirming the phenomenon of entanglement.

A particle accelerator is a machine capable of producing beams of ions or subatomic particles, including electrons, positrons, protons and antiprotons, with high kinetic energy. These machines are used for industrial, medical purposes and for the study of the structure of materials. At the level of theoretical experimentation, in particle physics their use allows to obtain high energy particle beams used to probe very small objects. In particular, the "Large Hadron Collider" (LHC) in Geneva, which came into operation on 10 September 2008, extends over a circumference of 27 kilometers and was designed to accelerate protons up to allow their study under conditions similar to those of the first moments of life of the Universe, immediately after the Big Bang.

Though absorbed in these commitments, Bell also found time to devote himself to the basic questions of quantum theory that had been forgotten. In 1964, about thirty years after the essay on the EPR paradox, Bell reopened the debate on this issue that seemed by now to be set aside and wrote an article for the journal "Physics" entitled "On the Einstein-Podolsky-Rosen paradox" (On the Einstein paradox -Podolsky-Rosen) in which he faced the problem of hidden variables.

Bell's question was whether quantum theory needed hidden variables, as argued by Einstein, in order to be reassuringly placed in the category of deterministic phenomena. He believed that this was essential to get quantum physics out of the impasse of non-locality, that is, from the possibility that two particles could communicate with each other in a non-local way, influencing each other without limits of space and time.

As we have already seen Einstein, despite being one of the founders of quantum theory, he doubted its completeness. From his point of view, a more complete theory would have had to predict the existence of elements of reality, or hidden variables, to preserve the concept of locality. Moreover, the non-locality would have been in contrast with his theory of relativity. Indeed, the theory of relativity excludes that the effects of a physical action can be observed instantaneously at a distance. This would have led to the transmission of information faster than the speed of light. Instead, the Einsteinian theory of relativity has among its foundations a very precise speed of light, equal to 299,792.458 kilometers per second. It is a parameter which, if not taken into consideration, would cause the whole theory to collapse.

Unfortunately, at the time when Einstein stated his objections, we were still working in the purely theoretical field, given the impossibility of carrying out checks due to the inadequacy of the experimental hardware. Consequently, Einstein's "local realism", valid on the theoretical level but contrasted with equally theoretically valid hypotheses, remained confined to the level of a topic of philosophical debate.

The unlocking of this impasse came when, in 1964, John Bell put the question back on the table by posing the problem in terms that can be verified by experiments. These had to provide for the measurement of physical properties of particles distant from each other, but linked to each other (correlated) by quantum effects.

Bell formulated the famous "Bell inequality" on the merits. In this theory he demonstrated that no addition of hidden variables, as presented in the EPR paradox, would be able to reproduce all the properties predicted by quantum mechanics, in particular those concerning entanglement.

Bell's inequality highlighted an alternative: either a quantum theory that fell under the umbrella of the Einsteinian locality was correct thanks to the contribution of hidden variables, or the quantum mechanics of Bohr and the Copenhagen School was correct, with its implications of non-locality. It excluded, of course, that both theories could be correct.

Actually Bell developed his theory starting from the assumption that there was indeed the possibility of imagining the integration of quantum theory with hidden variables, as Einstein had suggested.

With this he was in agreement with other studies conducted by David Bohm, which we will discuss shortly, according to which the hidden variables were vectors of information. That is, he suggested that the particles were already equipped with a set of instructions that told them in advance what their behavior should be under all circumstances.

Bell's inequality had the advantage of allowing an experimental verification by measuring the polarization of photons. The experiments carried out subsequently were the "smoking gun" against local realism and showed that the "gruesome actions at a distance", envisaged by the EPR paradox, are firmly part of reality.

Above all, these experiments undoubtedly demonstrate that, even in the hypothesis of the existence of hidden variables, it would still be impossible to deny the principle of non-locality.

In his book "Information. The new language of science" Hans Christian von Baeyer, an American academic, writes: "Contrary to Einstein's intuition, not only does God play dice, but he seems to be the only one we can trust to have a honest game.."

The experiments confirming non-locality.

In 1969 Bell received a letter from the young Berkeley physicist John Clauser telling him that he had devised an experiment to test his own theory of inequality. He wasn't the only one to perform these kinds of experiments. Verification of experimental results tended to ascertain the local or non-local nature of the real physical world. The advancement of technology over the years has led to significant improvements in efficiency, but in any case all the experiments confirmed the quantum theory of non-locality and provided results that could not be explained even by adding hidden variables.

In 1972 John Clauser, in collaboration with Stuart Freedman, used calcium atoms to obtain, with the double quantum leap of an electron, a pair of correlated photons. By carrying out a series of measurements for about 200 hours, Clauser and Fredman obtained confirmation of the phenomenon of entanglement and non-locality. This result puzzled Clauser himself, who tended to demonstrate the inconsistency of quantum theory. Perhaps this is why the experiment did not receive the attention it deserved.

Different was the case of the experiments carried out by Alain Aspect, (fig. 49) born in 1947 in Agen, a small French village not far from Bordeaux.

Contrary to Clauser, Aspect understood that quantum theory was achieving great scientific success at the theoretical level, and that the conflict between locality and non-locality suggested by Bell's inequality could be solved by experiments that would defeat the locality thesis; Aspect was confident that he could demonstrate quantum theory and its non-local effects.

In the mid-1970s, when he was still a doctoral student, he proposed a real experiment, which relied on two single but correlated photons emitted together by the same atomic source. In his experiment, the two photons would have traveled in opposite directions to each other and would have been controlled by two detectors placed at a distance.

The idea was very original. Bell, when asked, gave his approval; but with reservations. First he wanted to verify that the young Frenchman already had a fixed salary. In fact, Bell believed that the idea of Aspect was so bizarre that it could have compromised his career.

After a long preparatory work Aspect and his collaborators, Jean Dalibard and Gérard Roger, two researchers of the "Institut d'Optique Graduate School", carried out the experiment three times between 1981 and 1982. Their commitment did not decrease due to economic difficulties. Among other things, they handcrafted polarizers which, thanks to their fast rotation, could have given an answer to the question of location. These polarizers are currently exhibited in the historical showcase of the Institut d'Optique where the experiments took place in Orsay's dependency. The headquarters of the Institut, in turn, is located in Palaiseau, on the campus of the Paris-Saclay University.

Aspect's work was awarded; he and his team found the best possible confirmation of non-locality in quantum mechanics and flatly denied any possible local hidden variable theory.

The consequences of the experiments went far beyond demonstrating the validity of quantum mechanics, but highlighted the fundamental role played by entanglement. As we have seen, in the entangled state the observation of the state of one photon allows us to deduce the state of the other.

Figure 49 - Alain Aspect. From the beginning of the 1980s he carried out the experiment proposed by John Bell several times to confirm the EPR paradox. The experiment, on the other hand, confirmed the theory of entanglement of correlated particles.

Aspect's experiments have had enormous resonance and have given way to a long succession of theoretical and experimental investigations on entanglement. It can be said that Alain Aspect's experiments marked the beginning of Quantum Computer Science.

In 2013 Aspect received the prestigious Balzan Prize with the following motivation:

> "For his cutting-edge experiments that confirmed Quantum Mechanics, in contrast to the theories with local hidden variables. His work paved the way for the experimental control of entangled quantum states, an essential element of Quantum Computing."

Among the subsequent experiments it is useful to remember that of Wolfgang Tittel. This scholar earned his PhD from the University of Geneva, worked as a postdoc at the University of Aarhus and the University of Geneva and in 2006 entered the University of Calgary (Canada) as president of industrial research. He has been engaged in groundbreaking experiments since the beginning of quantum communication studies applied to cryptography.

In 1988 together with the "Geneva Group" he carried out an experiment with which he demonstrated that distance does not affect entanglement. The light was sent through fiber optic cables over distances of several kilometers before being analyzed, without the results varying.

Ten years later, in 1998, Gregor Weihs and his Innsbruck team headed by Anton Zeilinger carried out an experiment that improved upon Aspect's 1982 one, confirming the same results.

Unfortunately, Bell died unexpectedly in 1990 of a brain hemorrhage without being able to receive the Nobel Prize for which he had just received the nomination.

Part Three. The prospects.

Everything is "one thing".

If God created the world, we cannot say that he took great care to facilitate understanding.
(Albert Einstein)

Alain Aspect's experiments represented the definitive confirmation of the predictions of quantum mechanics, and in particular of non-determinism and non-locality.

In the context of general acceptance of these principles there was still some scientist who distinguished himself as a nonconformist and did not want to give up the attempt to frame quantum mechanics in a deterministic context. I refer to David Bohm. However, in his desire to rationalize the non-rationalizable, he ended up elaborating an even more romantic version, if possible, of the orthodox quantum one, and fully deserved the placement in the Olympus of the philosophers of science.

The story I tell in this third part is that of an extremely honest and determined man in his daring scientific theories, who did not hesitate to transpose his theories into social behavior. This earned him powerful antipathies whenever he found himself operating in conformist and politically bigoted contexts, such as those typical of his time.

David Bohm

David Bohm, an American physicist and philosopher, was born on December 20, 1917 in Wilkes-Barre, a city in the United States of America located in the center of the Wyoming Valley, in Luzerne County, in northeastern Pennsylvania.

He is mainly known for having come up with an interpretation of quantum mechanics in which he develops Louis de Broglie's concept of the "pilot wave" and introduces the concept of quantum potential.

His parents were of Jewish origin. His father, Samuel Bohm, was a Hungarian emigrant and his mother was from Lithuania.

The father, who owned a furniture shop, was also an assistant to the local rabbi. He himself provided for the youth education of his son David. At the age of 17, in 1930, David began attending Pennsylvania State College, where he made his first acquaintance with quantum mechanics and was attracted to it, making it the subject of his studies until he graduated in physics in 1939.

David continued his studies under the guidance of Robert Oppenheimer. Remember that the fame of this scientist is mainly linked to the construction of the first atomic bomb as part of the Manhattan project. Subsequently, Oppenheimer had a crisis of conscience and refused to work for the hydrogen bomb.

Perhaps it was Oppenheimer's teachings that helped further develop Bohm's strong moral sense of justice.

After continuing his studies at CalTech (California Institute of Technology) in Pasadena, he enrolled at UC (University of California) in Berkeley, where he obtained his doctorate in 1943. Bohm practiced his activity as a theoretical physicist in the most important American universities.

From 1947 to 1951 he taught at Princeton University while dedicating himself to the study of plasmas.

Plasma is a highly ionized gas and, on the whole, electrically neutral: concretely, a system made up of free positive and negative ions, with equal concentrations.

The term "ionized" indicates that a significantly large fraction of electrons have been stripped from their respective atoms.

Plasma is considered to be the fourth state of matter, which stands out and adds to the solid, liquid and gas. In this sense, it reproduces the idea of the four elements (fire, earth, air and water), which historically goes back to the Greek philosopher Empedocles.

Plasma was identified by Sir William Crookes in 1879, but the name "plasma" is due to Irving Langmuir who coined it in 1928.

Until the end of the last century, studies on plasmas remained mostly a curiosity, with very few exceptions, among which Nikola Tesla's on plasma discharges, plasma lamps and cold plasma used in the production of ozone can be remembered. After 1920 Irving Langmuir studied the interaction of plasmas with the walls of the container in which they were formed, obtaining the Nobel Prize in 1932.

A more concrete interest in plasmas developed when the Geneva Atoms for Peace Conference initiated studies on the peaceful exploitation of nuclear fusion.

In his study on plasmas Bohm immediately noticed the peculiarity, typical of electrons when they become part of a plasma, of losing their "individualism" to begin behaving as part of a larger and interconnected reality, giving the impression of being animated by an intelligence coordinating their activity.

He soon discovered, together with his colleague Yakir Aharonov, the Aharonov-Bohm Effect, a phenomenon of quantum mechanics in which a charged particle is affected by electromagnetic fields in regions where these fields are zero.

In 1947 Bohm studied the behavior of electrons in metals. On the basis of what he observed for plasmas, he noticed that even in metals the individual movements of the single electrons are able to produce organized "collective" effects.

In his philosophical interpretation Bohm came to identify the behavior of electron masses with that of communities of human beings. He established a connection between the coordinated behaviors of individual electrons in metals and those of individuals in their social contexts. According to Bohm, the typical behavior of plasmas was a valid metaphor for organized human behavior. He developed the conviction that the greatest analogies were those between the "plasma societies" and human societies organized according to Marxist theories.

These suggestions reinforced Bohm's early Marxist ideals. In later years he decisively abandoned these ideas for their materialistic foundation, and preferred to devote himself to a more spiritual view of the cosmos.

The results of his studies on the behavior of electrons in metals convinced his old teacher, Oppenheimer, to include him among the scientists called to work at Los Alamos, on the Manhattan Project, a research and development program in the military field that led to the creation of the first atomic bombs during the Second World War. Unfortunately, Bohm had a political background that was not very reassuring for the time and General Leslie Groves opposed Bohm's entry into the project, due to the ideas he professed in his youth.

Figure 50 - Physicist David Bohm proposed a fascinating holistic interpretation of quantum mechanics, which countered the orthodox view known as the "Copenhagen Interpretation".

Indeed, between the end of 1940 and the mid-1950s the maximum phase of the so-called "witch hunt" or McCarthyism was being experienced. This was a widespread political attitude in the United States of America, characterized by an exasperated climate of suspicion and persecutory behavior towards people or groups deemed subversive. It was named after Senator J.R. McCarthy, director of the Commission for the Suppression of Anti-American Activities.

Thousands of people were subjected to surveillance and many suffered personal attacks, often based on unsubstantiated allegations.

Particular attention was paid to government officials, entertainers and men of culture and leading figures of the industrial middle class who were branded as "Communists committed to undermining the foundations of American society".

In this climate, in 1948 Bohm was called to testify due to his accession to the Communist Party which took place in 1942 and ended after only nine months, but he refused to appear appealing to the Fifth Amendment, that is, the right not to answer.

After a year, he was again invited to appear before the Grand Jury, but once again he refused, appealing to the same constitutional provision. But now the attention was focused on him. In 1949 he was arrested for contempt of court and served a few days in prison, until he was released on bail.

In 1950 he was acquitted but, in the meantime, he had lost his job at Princeton University, which had fired him due to pressure from his sponsors.

By now Bohm was on the banned list of researchers and not even Einstein, who would have wanted him as an assistant, was able to offer him the opportunity. He was forced to emigrate to Brazil at the University of Sao Paulo, but he was still an American citizen, so the US embassy withdrew his passport fearing that he might go to the Soviet Union.

To avoid complete isolation, Bohm took Brazilian nationality. This allowed him, in 1955, to leave Brazil and go to Israel for two years, at the Technion Institute in Haifa. From there he moved for four years to the University of Bristol in Great Britain and finally, in 1961, he settled permanently in London where he was appointed professor of theoretical physics at Birkbeck College.

In 1951 Bohm published "Quantum Theory", a text on quantum physics in which he analyzed the interpretation of quantum theory and, taking up some ideas from the EPR thought experiment, he placed himself among the critics of the Copenhagen interpretation. In fact, Bohm highlighted the unclear aspects of the theory, and contested the impossibility of extending randomness and determinism to the objects of quantum mechanics. Bohm, like Einstein, was deeply troubled by the position of the scientific community which now "supinely" accepted the reality of indeterminacy and non-locality.

Bohm saw it as his mission to reintroduce a form of determinism that worked even in the extremely small. He felt gratified by giving Einstein a copy of his book..

The quantum potential.

Heisenberg's principle states that it is not possible to establish what the trajectory of a particle, such as an electron, is, since it would be necessary to know its speed and position at the same time. As already mentioned, the only possibility is to calculate a wave function, within which the particle occupies a probabilistic position. When a measurement is made, the particle "collapses" and appears in a precise place, but between one measurement and another it "disappears" in a cloud of probability.

This did not please Bohm; he rejected the notion that the probabilistic function could be the best possible representation of a particle. He therefore decided to investigate the way in which an electron establishes the trajectory it will have to travel. Put simply, he tried to reintroduce the concept of causality, that is, to identify a precise mechanism capable of guiding particles. In this way the particles would be freed from the indeterminate and probabilistic context.

Bohm started from Heisenberg's uncertainty principle and reworked the formula, adding a new parameter, the "quantum potential". With this he created a new quantum mathematics, "Bohmian mechanics." This theory is still highly debated for its metaphysical implications.

The "Bohmian mechanics" proposes that the particle does not move randomly, but responds to a force, the" quantum potential. "This force provides the particle with information through instantaneous, ie non-local, connections.

To operate non-local, the quantum potential must retain its intensity regardless of distance. That is, the quantum potential must operate in an opposite way to that of all other electromagnetic forces. As is known, the other electromagnetic forces decrease due to the inverse of the square of the distance.

According to Bohm, the quantum potential drives the electron in a way that is neither random nor mechanical, but intelligent. In other words, Bohm argues for the existence of an information field, thanks to which the particle is not subject to the mysterious whims of chance but is constantly informed about the surrounding environment, obtaining information from the same environment. Evidently, Bohm suggests a metaphysical environment, without space and without time.

Therefore, when the particles move in their path (and remember that the particles are in constant agitation) they do not do it randomly but respond to the information packet provided by the quantum potential. In this way Bohm restores a deterministic behavior for elementary particles, which however is due to an absolutely mysterious force.

Bohm represents this concept with one of his famous metaphors, that of the ship.

A ship (fig. 51) is capable of moving thanks to the power of its engines, but it is guided by navigation instruments, without which it would never reach a port and would risk crashing into a shoal every moment.

Metaphors by David Bohm.
The Ship
Quantum potential represents
the information that
drives the ship
The engines represent the force that drives the ship

Figure 51 - In the metaphor of the ship, Bohm represents the particle as a ship pushed by the "brute force" of its engines but guided by the intelligence of the instrumentation, that is, by the quantum potential that draws on global information and transmits it to the ship.

Metaphors by David Bohm. The ship

A The quantum potential represents the information that drives the ship

B The engines represent the force that propels the ship.

The engines represent the power, the certainties of classical physics that make the ship move in the sea. Instead, navigation instruments, from the ancient compass to the modern radar, are the information that determines the safe course and represent the quantum potential.

A small radar signal, not too strong, is enough to guide the ship. According to Bohm, the quantum potential does not depend on intensity but on form. It is an energy that, although modest, is very well informed and manages to give shape to the great energy of the engines, which in itself would be "raw" and "formless"

Many concrete elements suggest that quantum potential is not just a theoretical abstraction of Bohm. The existence of an immense sea of energy that pervades the entire universe is predicted by the same classical quantum theory. This near-infinite deposit would be the "quantum vacuum", also often referred to as the "zero-point field" or the "Casimir effect". This last definition derives from the fact that the existence of the "quantum vacuum" was demonstrated by the Dutch physicist Hendrik Casimir as early as 1948.

The Casimir effect consists in the attraction that is created between two bodies located in the void, that is, where there should be no reason for attraction. However, since the attraction is there, it means that the void is not really empty. In fact, the presence of a force, called the zero point field, is assumed. This field arises from the vacuum energy, determined by virtual particles that are created continuously as a result of quantum fluctuations, as foreseen by the Heisenberg uncertainty principle.

Therefore, the vacuum has an "underlying energy" or "transcendent the reality of matter", which extends into every "empty" area of the universe. The density of its energy has been estimated to be equal to immense values, which establish a disproportionately unbalanced relationship with that of the cosmos we know and which is presented in the form of matter.

If we take the image of the atom, we remember that it is considered to be composed almost entirely of void. Well, the Casimir effect predicts that that void is not empty, but energy. From this we understand how the energy of the vacuum is immensely greater than that of the "non-vacuum". If we allow all this energy the ability to express itself in the form of a quantum potential, we realize the power of the concept developed by Bohm.

The role of the observer.

Quantum mechanics predicts that cosmic reality is the result of two factors: a "brute" force inherent in matter, and a "guide system" that provides the matter with instructions capable of making its force work intelligently.

The first factor, brute force, characterizes the macroscopic, or medium-sized, level in which we live and which we interpret according to the laws of Newtonian physics and Einstein's relativity. At this level, the role of the observer has no influence. In fact it is very difficult that, by observing a phenomenon, we can change its development. We will not be able to move any chairs by looking at it. At the same time, no chair will move in two different paths when we are not looking at it.

The second factor, that is, the guidance system, provides that the observer is part of the phenomenon itself and that it affects its development simply by observing it. As already mentioned. this happens in the case of photons which behave as particles when they are observed and as waves when they are not observed. At the subatomic level, the role of the observer takes on very particular characteristics. The observer becomes part of a single system that includes both himself and the observed object. Observer and observed become indistinguishable and exert a reciprocal influence that cannot be ignored.

John Wheeler was an American physicist, a charismatic figure in physics of the 1930s and 1940s; under his leadership many famous physicists were trained including Richard Feynman. Wheeler, among other things, coined the term "warmhole" which indicates the possible space tunnels or passages between the different regions of space-time.

Wheeler believes that the observer's involvement in the observed subatomic reality is even the most important aspect of quantum physics, to the point that he proposes to eliminate the term "observer" by replacing it with that of "participant". He expresses this belief in a famous quote:

> "In quantum mechanics nothing is more important than the fact that the concept understood as" something that is out of here "is destroyed.
>
> This is true even when the observer is at a safe distance, separated from the observed object with glass plates twenty centimeters thick. When he observes an object as tiny as an electron, the observer has to go past the glass: he has to "enter" to install the measuring device he has chosen. It is up to him to decide whether to measure the position or the momentum. The installation of the device to measure one of the two quantities prevents him from installing another device to measure the other quantity.
>
> Furthermore, the measurement changes the state of the electron. After the measurement, the universe is no longer the same. To describe what happened, the old word "observer" must be removed and replaced with the new word "participant". In a way, the universe is a participatory universe."

I have already dealt with this topic in the first part of the book, to which I refer: The universe exists because we look at it.

No fragmentation.

According to Bohm, the subatomic particles, thanks to the quantum potential, are immersed in a "field of form". From this field, the particles receive all the information necessary at any moment to decide how to move.

But, since the macroscopic level is composed of an almost infinite number of subatomic particles, every living being is subject to a field of form which, at every instant, suggests to every particle of its body the coordinates of its microscopic movements.

The difference between the two factors (macroscopic and microscopic) is that the former needs space and time to take place, that is, it has local characteristics, while the microscopic factor, that is the guide ṣystem, needs neither space nor time. and therefore it exists in a non-local dimension. The guidance system is the expression of an "infinity" located outside of space and time. The guiding system does not produce information by receiving it from a specific place, but possesses it in a "universal whole", a kind of prespace that contains the consciousness of the universe.

In the context of Bohmian mechanics the particles of a system, at whatever distance they can be found, are connected in a nonlocal way through the quantum potential. The most immediate consequence is the realization that there are no separate and independent objects, but the whole is interconnected to form a single thing.

Moreover, as we have already seen, there is no whole that is "other" from us and that we can observe from the outside, but we ourselves are an active part of the whole. Quantum mechanics manifests a totally interconnected universe. It is a reality that we cannot break down into elementary parts with independent life, to the point that not even ourselves, observing it, can call ourselves out, establishing that "the universe is out there and we are here". By studying the intimacy of matter, we come to understand that the particles of which it is composed do not represent the fundamental building blocks in the sense intended by Democritus. Particles are pure idealizations, study hypotheses capable of providing ideal representations of a reality that is impossible to represent. In fact, particles are conventional abstractions.

In reality there is a cosmic interconnection made up of probabilistic vibrating energies, which cannot be placed exactly in space as we understand it. On the merits, Bohm expresses himself in this way in the essay written together with Basil Hiley for the journal "Foundation of Physics", published in 1975 with the title "On the intuitive understanding of nonlocality as implied by quantum theory". (On the intuitive understanding of non-locality as implicit in quantum theory):

> "We have overturned the usual classical conception according to which the independent" elementary parts "of the world are the fundamental reality, and the various systems are only particular and contingent forms and arrangements of these parts. Indeed, we say that the fundamental reality is the inseparable interconnection quantum of the whole universe and that the parts that have a relatively independent behavior are only particular and contingent forms within this whole ".

As if to say that King Milinda's chariot is not the completely random sum of wheels, axle and horses, as claimed by materialism, but that these elements are, instead, occasional parts of a reality that is the chariot.

Whatever form the axis takes, whatever number of spokes the wheels have, whatever the color or number of horses, is completely irrelevant to the triumph of the transcendent chariot project.

At the subatomic level, the objects in which solid matter materialize dissolve in probability waves interconnected into a whole.

There is no fragmentation, but only a unicum teeming with vibrations. If we wear quantum physics glasses we can no longer see the universe as a collection of different objects, from pollen grains to galaxies, well defined and distinct from each other. What appears to us is an immense network of related energies in a whole that also includes ourselves.

In the essay "S-matrix interpretation of Quantum Theory" the American physicist Henry Pierce Stapp states:

> "The world appears as a complicated web of events, in which different kinds of interconnections alternate, overlap and combine, determining the structure of the whole".

Figure 52 - Bohm's metaphor on quantum superposition.
Metaphors by David Bohm. The fish.
The same fish, when viewed from different angles, can give the impression that they are two fish.

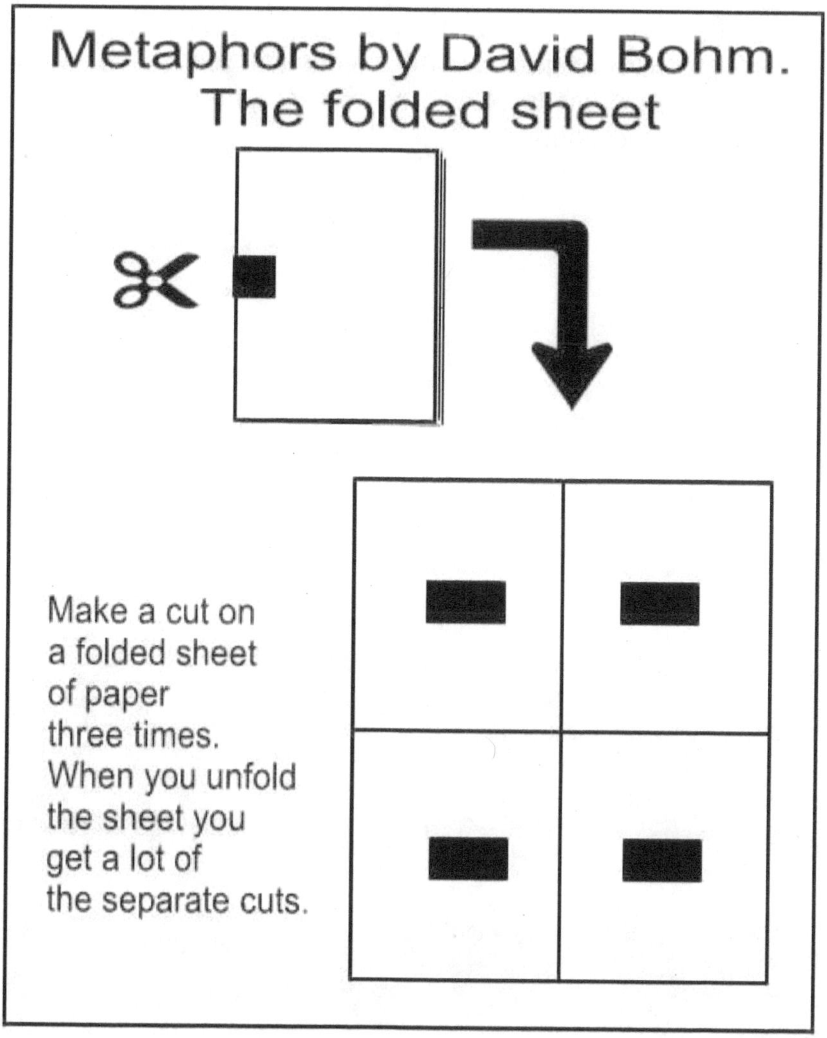

Figure 53 - The many holes that appear on the unfolded sheet are, in reality, the result of a single cut.

Metaphors by David Bohm. The folded sheet.

Make a cut on a sheet of paper folded three times.

When you unfold the sheet you get many separate cuts.

The metaphor of the fish being photographed from different angles (fig. 52) superbly describes the concept according to which, although there are apparently different manifestations of an object, such as to lead one to believe that it is several distinct objects, in fact it is a object only

"Implied Order" and "Explicit Order".

"On a very deep level matter and consciousness are completely inseparable and interconnected, just as in a video game the player and the screen are united by participation in a common process. In this view, mind and matter are two aspects of a single whole and are no more separable than form and content are. At very deep levels, the consciousness of humanity is one.
(David Bohm)

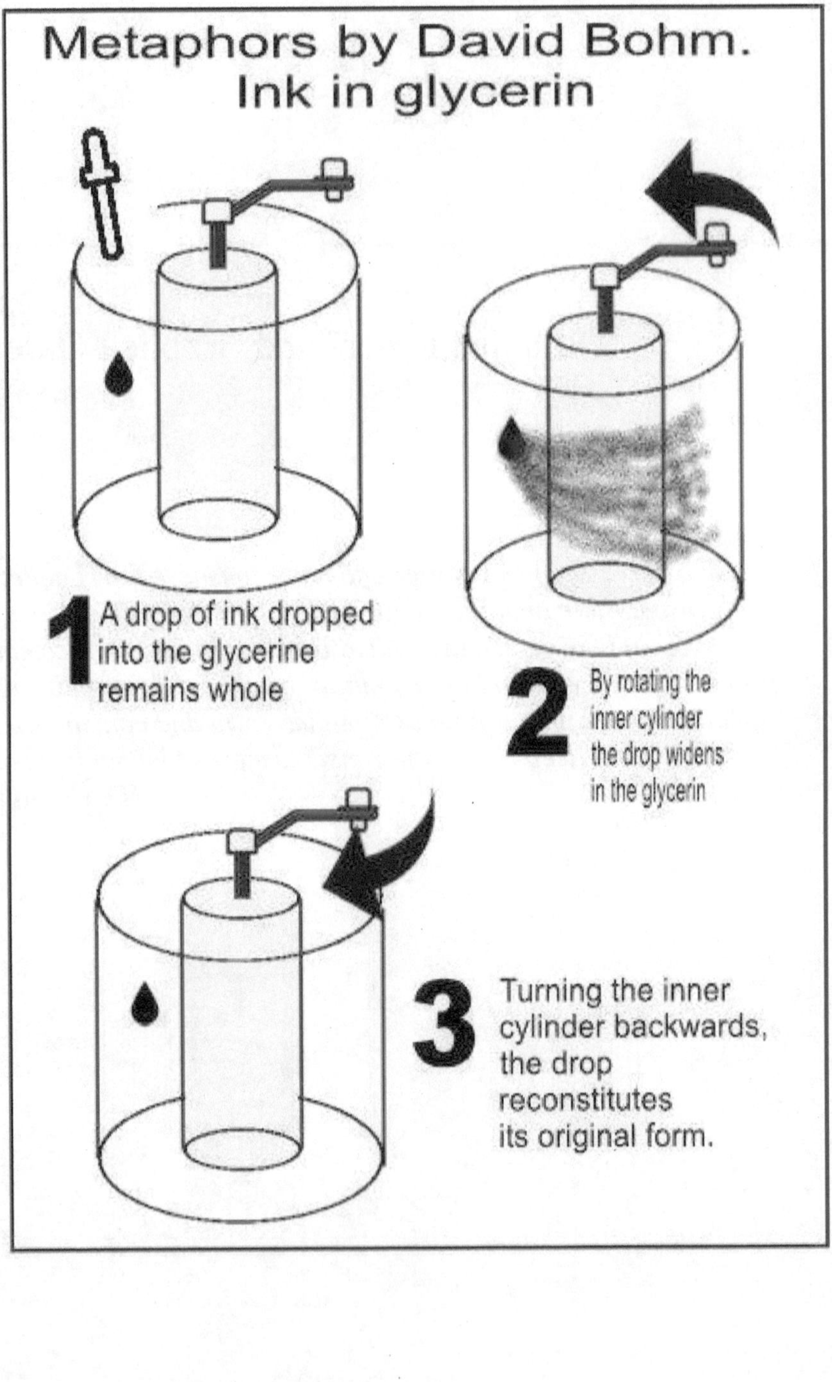
Metaphors by David Bohm.
Ink in glycerin
1 A drop of ink dropped into the glycerine remains whole
2 By rotating the inner cylinder the drop widens in the glycerin
3 Turning the inner cylinder backwards, the drop reconstitutes its original form.

Figure 54 - David Bohm's metaphor about a drop of ink dispersed in glycerin.

Metaphors by David Bohm. The ink in the glycerin

1 A drop of ink dropped into the glycerin remains whole

2 By turning the inner cylinder, the drop spreads into the glycerin

3 By turning the inner cylinder in the opposite direction, the drop restores its original shape..

This led him to theorize the existence of a transcendent order of time and space, inherent in the deepest folds of reality, which he called "implicate order" or "enfolded order". Bohm assumed that this order did not exist only in the world of elementary particles, but in the entire universe. The implicate order, typical of the most hidden part of the cosmos, is opposed to the "explicate order)". This is the order of things as we see them, or as they appear to us in our experience.

Bohm had the first intuitions of the existence of an implicated order when he dreamed of a universe composed of an infinite number of spherical mirrors, which were reflected in each other. He imagined matter as composed of atoms where each atom, like a mirror, reflects infinite reflections.

The metaphor of figure 53 illustrates what relationship may exist between the implicate order and the explicated order. Let's imagine folding a sheet of paper into three or more parts. With the scissors we open a single window at any point on the sheet. It will be enough for us to unfold the sheet to notice the existence of many windows, spatially separated from each other: this separation is completely illusory if we consider the folded sheet, but it is expressed to our senses as a sum of windows when we unfold the sheet. Evidently, the folded sheet represents the implied order, while the unfolded sheet represents the explicated order.

Bohm loved to elaborate metaphors to support his thesis. As regards the implicate and explicit order, he imagined an implicated dimension containing the information of reality in a form that is not intelligible to our senses, but capable of being expressed in forms and ways that are understandable to us, even if no longer corresponding to reality.

In the metaphor of the TV signal (fig. 55) the implicated order is composed of radio frequencies, invisible and incomprehensible to us, which extend into the ether and can be picked up by an antenna. When transferred to the receiver, the radio frequencies are expressed so that our eyes and ears can share them. Of course, what we see on the screen is only the transposition into shapes and colors of an otherwise indecipherable reality.

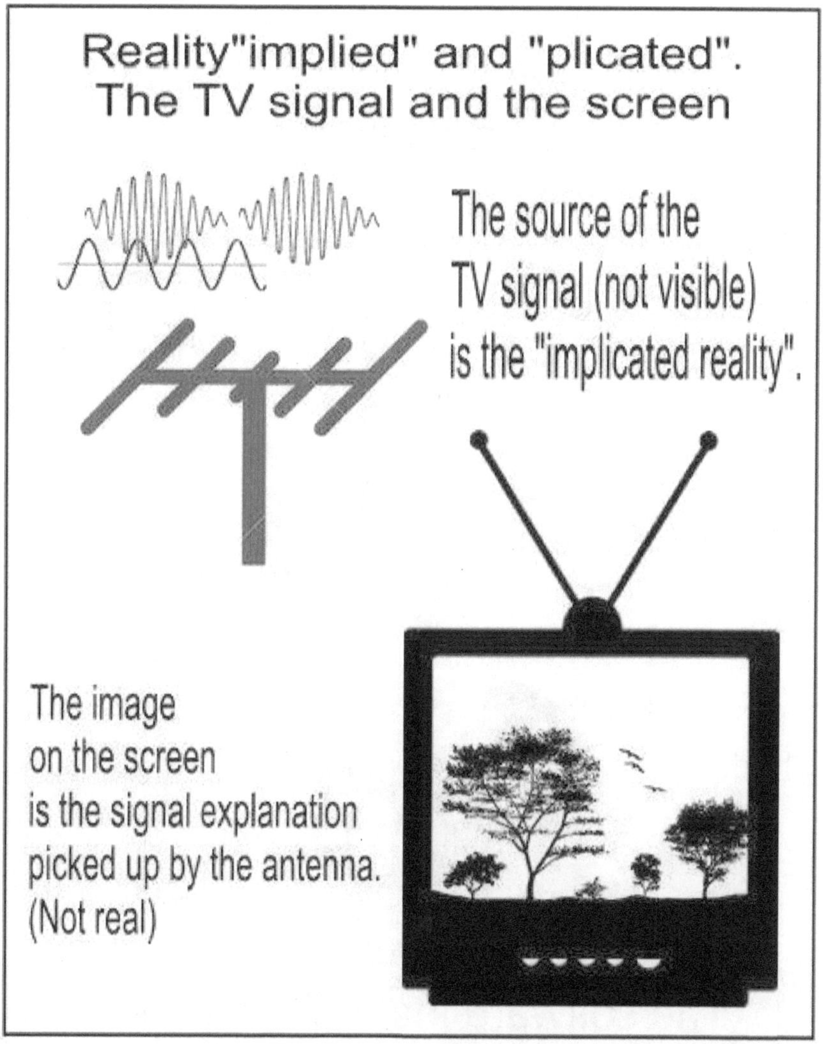

Figure 55 - The "reality explicated" is only an unreal interpretation of the signals received from an external source through the antenna.
Reality "implicated" and "explicated".
The TV signal and the screen
a - The source of the TV signal (not visible) is the "reality implied".
B - The image on the screen is the explanation of the signal picked up by the antenna. (Not real).

Reality"implied" and "explicated".
The video game
The events that
develop on the screen
are the
"reality explicated"
The software that guides events
is the"implicated reality"

Figure 56 - Explicit reality (what you see on the screen) is the representation, adapted to our senses, of the implicated reality, that is, of the software coming from the cartridge loaded in the console.

Reality "implicated" and "explicated". The videogame

a - The events that develop on the screen are the "reality explicated"

b - The software that guides the events is the "reality involved".

Even more fitting, if possible, is the metaphor of the videogame (fig. 56) where an intelligence alien to our environment and our conscience, that is the software coming from the cartridge inserted in the console, creates for us the illusion of a fantastic world with which we can interact as if we were really immersed in it while, in reality, it is pure illusion.

Another metaphor provides that the two orders can be illustrated by setting up the device in figure 54 which was explained by Bohm himself during an interview with the oubliettemagazine.com web-site:

> "When I was thinking about the universe involved, a science program at the BBC showed a device that illustrates the concept well. It consists of two concentric glass cylinders. Between them is a viscous fluid, such as glycerin. If a drop of insoluble ink is put into the glycerin and the outer cylinder is rotated slowly, the drop of dye will be stretched into a trail. Eventually, the trail becomes so diffuse that it becomes almost invisible. At that moment there seems to be no order present. However, if you slowly rotate the cylinder backwards, the glycerin returns to its original shape and suddenly the drop of ink becomes visible again. The drop of ink that was dispersed in the glycerin is reconstituted thanks to the reverse rotation. "

We have two concentric glass cylinders, one of which is able to rotate slowly. The space between the two cylinders is filled with a viscous liquid such as glycerin. When a cylinder is rotated, the nearby fluid is dragged along with it. But suppose the movable cylinder is now rotated in the reverse direction. If the fluid is very viscous like glycerin, and the cylinder is not rotated too fast, then the fluid element will retrace its steps exactly and reconstitute the original ink drop.

This metaphor teaches that there is a hidden order even in apparent chaos. When the shape of the ink drop dissolves until it disappears, its order is not destroyed but remains implicit in the glycerin. This dissolved drop represents the implicate order, while the whole drop represents the explicated order, that is, reality as it appears to our senses. The rotation of the cylinder plays a decisive role and represents the dynamics that unify the two orders. In the Bohmian elaboration this dynamic is called "holomovement".

A giant cosmic hologram.

The term hologram generally identifies three-dimensional images, and in particular stereoscopic images, that is, those that appear with different perspectives depending on the point of observation. Therefore, a hologram is a three-dimensional representation of reality, and could be compared to a normal photograph if this were not two-dimensional. By photographing an object with classical methods, you can then see it exactly from the point of view of the lens used to take the photo. If you film a person in front you will always see her face but you will never be able to see the back of his neck. Conversely, a hologram will give you the three-dimensional image of the scene shot, where objects and people will appear as placed in the center of a stage.

Holograms are obtained with special techniques. Registration takes place via a beam of laser light sent both towards the object to be reproduced and towards a plate of sensitive material.

Thanks to a game of mirrors, the light coming from the source interferes with that reflected from the object. As a result, lines are formed on the plate, called interference fringes. In simpler words, the image of the photographed object does not appear on the plate, but a tangle of lines that contain information on three-dimensionality.

Of course, the recorded image can be reproduced with remarkable precision, simply by illuminating the plate with another laser beam.

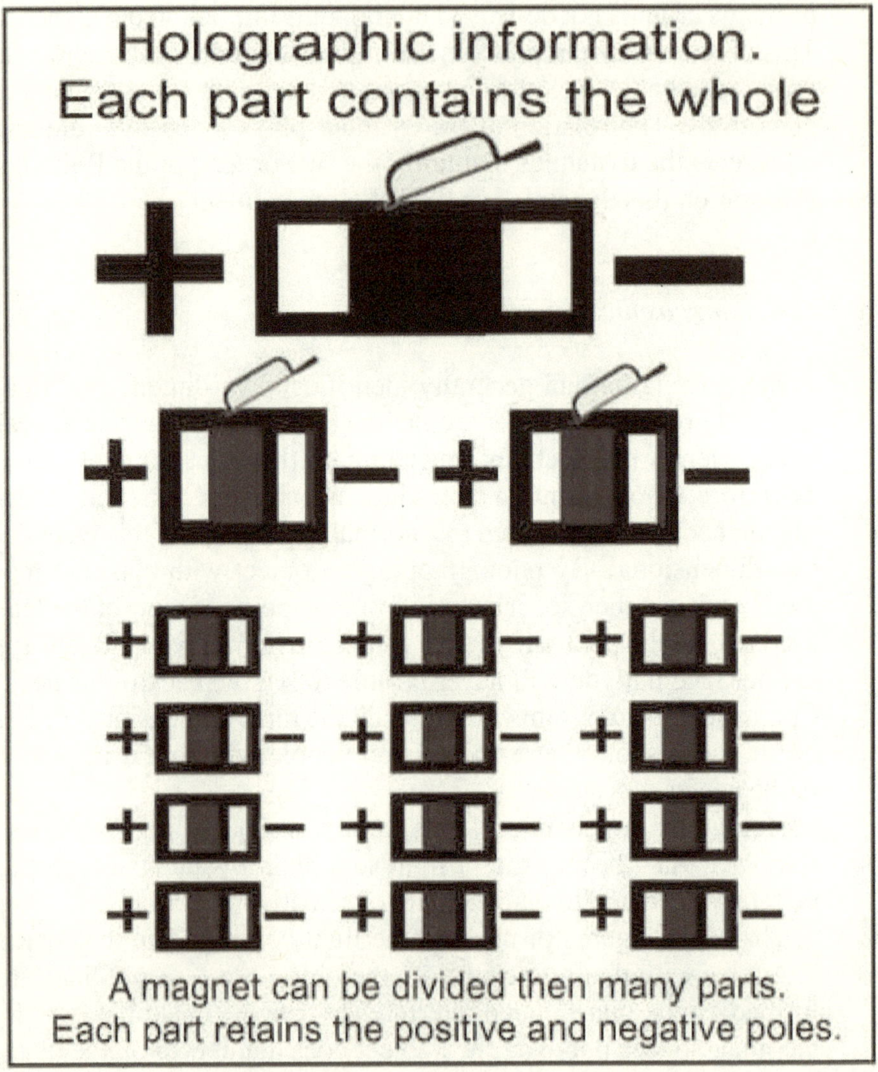
Holographic information.
Each part contains the whole
A magnet can be divided then many parts.
Each part retains the positive and negative poles.

Figure 57 - A magnet can represent a holographic reality. If the magnet is split into any number of parts, each part retains the characteristics of the whole.

Holographic information. Each part contains the whole

A magnet can be divided into many parts. Each part retains the positive and negative poles.

This, in fact, manages to decode the information by reconstructing the three-dimensional image of the object being shot, which appears to the viewer as if it were physically present. Thanks to technical improvements and developments, holograms can now also be visible in "white light" or sunlight.

Today there are also printed holograms, in which the diffraction grating is reproduced on a transparent plastic support, placed in turn on a silver layer. We can see similar holograms, for example, on some credit cards.

A hologram has a very interesting peculiarity, which clearly distinguishes it from a normal photo in which the information is crystallized on the paper. If you cut a photograph in two parts and throw away one part, the information that was contained in it can be considered permanently lost. You will never be able to reconstruct, from the surviving half, the content of the missing one, if not by intuitive approximation.

On the contrary, a hologram can be divided into as many parts as you want, and each part will keep the total information.

Bohm suggests the metaphor of the magnet (fig. 57). This object, commonly called a magnet, is a metallic body that generates a magnetic field invisible to the human eye, and produces the universally known effects: it can attract or repel other magnets, in addition to ferrous materials. The magnet has two poles, called north and south. If the magnet is allowed to rotate freely, the north pole is the one facing north. Obviously, at the opposite end is the south pole.

One might think that, by sawing the magnet in the middle, two parts are obtained, one with only the north pole and the other with only the south pole. It is not so. Each of the two parts immediately reconstitutes the missing pole. You can divide a magnet into as many parts as you like; in any case, however small it is, each part will be equipped with a north pole and a south pole, that is, each part will keep intact all the information of the original magnet from which it was obtained.

By cutting a hologram in two, ten, one hundred or one thousand parts, the same thing happens. Each part keeps all the information of the matrix from which it comes.

Moreover, a hologram has an incredible storage capacity from a quantitative point of view: just change the angle with which two laser beams engrave the film to accumulate millions of billions of information in a single cubic centimeter of space.

The analogy with the "implicated universe" is evident: here too each part is interconnected with all the others forming a single thing, and in this network of connections even the smallest part, like a single particle, contains the global information. .

At the same moment in which the information of the entire cosmos can be recognized in the smallest particle, the Bohmian theory assumes holographic characteristics and the universe, understood as an implicit and explicit order, can be imagined as an immense hologram.

The holographic universe.

The most astonishing thing about the holographic model was that it suddenly made sense of a wide range of incomprehensible phenomena, such as to be considered outside the confines of science. These include telepathy, precognition, mystical feelings of unity with the universe and even psychokinesis, or the ability of the mind to move material objects without anyone touching them."

(Michael Talbot, "All is One")

Bohm's intuitions soon took on a precise direction, that of a universe imagined as a gigantic hologram. It was necessary, however, to necessarily take into account the fact that the universe is not a static but a dynamic reality, an immense succession of vibrating energies and agitated matter; in fact, nothing in the subatomic universe is static, as we have seen when describing the atom.

Hence, Bohm coined the new term holomovement. The universe, in the Bohmian conception, is an immense holomovement in which every part, even the tiniest, contains all of history and cosmic intelligence as they unfold in time and space.

Precisely because the implicated universe implies within itself all the physical universes and their becoming to our senses, the resulting hologram does not even know the concept of stillness but constitutes an eternal movement of "implication" and "explication".

The implicated order contains in itself every time, past, present and future, and every place. Time and place, which in the explicit order (that is, in our experience and understanding) unfold in separate realities, are one in the implicate order.

The holomovement makes it possible for the implicate order to contain all the diversities of the explicate order. In an interview, Bohm explains the holomovement with these words:

> "The same implicit order has been detected in the complex movement of electromagnetic fields in the form of light waves. Such a movement of light waves is present everywhere and encompasses the entire universe of space and time in each region. This enclosing and unfolding takes place not only in the movement of the electromagnetic field but also in other fields that obey the laws of quantum mechanics that imply properties of discontinuity and non-locality. The totality of the closing and opening movement can go immensely beyond what was detected in the course of our observations. We call this totality "holomovement".

In the same interview Bohm says again:

> "I propose something like this: Imagine a sea of infinite energy filling empty space, with waves moving through it, gathering occasionally and producing an intense pulsation. Let's say that a particular impulse collects and expands, creating our universe of space-time and matter. More of these impulses may occur. To us, this impulse looks like a Big Bang; in a larger context, it is a small wave. Everything emerges from the deployment in the holomovement, then folds back into the implicit order. I call the process of folding "implicating" and that of unfolding "explaining". The implicate and the explicate are one flowing and undivided thing."

Bohm then completes his vision of the universe in an exquisitely mystical way by imagining, as in a matryoshka, very long theories of implicated realities underlying others, until he reaches a "super-implicated order", that is a "cosmic mind" that acts by experimenting creatively herself in the higher orders. Every living being consciously participates in the whole by giving a meaning to the whole. Without the consciousness of living beings, the universe would be incomplete. Thanks to the participation of every living being, the cosmic Mind matures, in the explicit order, awareness and knowledge of itself.

Karl Pribram

Karl Pribram was born in Vienna in 1919 and returned his life to heaven in 2015. He was a neurosurgeon, professor of psychiatry and psychology at various American universities, including Stanford and Georgetown universities.

Figure 58 - Karl Pribram, independently of Bohm, came to elaborate the concept of the holographic mind.

His most important studies concern the human brain and the nervous system: in particular he helped to define the nature of the limbic system, and the relationship between it and the frontal cortex.

It will seem strange that we talk about it in this context, dedicated for the vast majority to physicists or philosophers, but there is a very specific reason. Indeed, Pribram is mainly known for developing, in collaboration with David Bohm, the so-called "holonomic model of the brain" also known as the "Pribram and Bohm holographic model" or the "holographic paradigm".

With this model, Pribram theorized that the information, and therefore also the memories stored in our brain, are not "registered" in neurons, but are the result of interfering waveforms (patterns), thus explaining the ability of the brain to store a huge amount of information in a relatively small space.

Pribram believed that the memories were stored in the network of nerve impulses that weave throughout the brain, just like the laser beams that intersect on the film containing a holographic image and allow, depending on the angle, to reveal enormous amounts of information.

In his neurosurgical studies Pribram had come, like Bohm, to the conclusion that the functioning of memory was more understandable by considering it similar to a holographic reality. Naturally, both of them developed their convictions following absolutely different paths. Bohm formulated his theory following his studies in quantum physics, while Pribram became convinced of a holographic reality due to the many difficulties, sometimes insurmountable, in explaining some neurophysiological questions.

In the 1970s Pribram was exploring the hypothesis of a holographic reality, starting from the principle that our brain processes the signals received by the eyes in order to represent a mediated reality, similar to a hologram. However, he could not resolve a doubt: if it was a hologram, it was the hologram of what?

Since his son was a physicist, he thought of turning to him to compare some ideas, but the latter immediately advised him to contact David Bohm, who was certainly the most suitable person for him. Pribram did so and from their meeting an identity of views was born, solid as a rock, on the possibility that the universe was a holographic reality.

In his book "Holographic universe: The Revolutionary Theory of Reality" Michael Talbot describes the fruits of the meeting between the two scientists as follows:

> !Taken together, the theories of Bohm and Pribram provide a new, profound way of looking at the world. Our brains mathematically construct objective reality, interpreting frequencies that are ultimately projections from another dimension, a deeper order of existence beyond space and time: the brain is a hologram hidden in a holographic universe.
>
> Pribram argues that this synthesis made him aware that the objective world does not exist, or at least not in the sense we are used to believe. What exists "out there" is a vast ocean of waves and frequencies and reality only appears concrete to us because our brains are capable of transforming this fuzzy holographic form into the basic elements and familiar objects that make up our world. (...)
>
> We have two very different aspects of reality. We can see ourselves as material bodies moving through space, or we can see ourselves as an indistinct image of interference patterns hidden throughout the entire cosmic hologram. Bohm believes that this second point of view may even be the most correct one, because thinking of ourselves as a holographic brain-mind observing a holographic universe is again an abstraction, an attempt to separate two things that are fundamentally insuperable "

Doesn't reality exist?

Combining Pribram's holographic brain model with Bohm's theory of holomovement, a surprising result is obtained. If reality, in the concreteness we perceive, is illusion, and all that exists is nothing but a mass of vibrating frequencies and energies, what remains? If the brain is nothing more than a hologram that decodes the received frequencies, transforming them into sensory perceptions, does objective reality really exist?

The answer should be: no, "objective reality" does not exist, thus certifying what has always been maintained by Eastern philosophies, according to which the world is only an illusion.

Not only that: with this answer we also rehabilitate the theories, too hastily mocked, of many Western philosophers, such as George Berkeley, according to which all objects that we believe exist have no body, they are only ideas that our senses transmit to the brain. in which they are configured and covered with a false impression of materiality.

Pribram's holographic paradigm and Bohm's holomovement lead to conclusions that many contest, but many others believe to be the most accurate description of reality possible to date.

Where are we going.

"You have to understand the whole of life, not just a part of it. That's why you have to read, that's why you have to look at the skies, that's why you have to sing and dance, and write poetry, and suffer and understand, because all of this is life. "
(Jiddu Krishnamurti)

We are living in a privileged era in which a synchronicity of cosmic dimensions is guiding us towards unsuspected goals. Once we have reached them, it will happen that spirit and matter, finally clothed with the same dignity, will be able to involve all of humanity in knowledge and possibilities that today still appear incomprehensible and unattainable. This synchronicity is guiding us in the transition between the age of gray matter and that of the luminous spirit.

A first important step has already been achieved with the defeat of materialism, despite the fact that even today there are places, entities and people who do not renounce to place it as the primary assumption of their intellectual exercise, probably to protect positions of privilege or interest.

By now materialistic theories are totally outdated in the most advanced scientific fields, to the point that one wonders if those "randomly assembled machines" as we thought we were, do not instead possess a consciousness capable of modeling the universe that contains them.

In this context, a question that, until the last century, was confined to the nebulous realm of metaphysics, becomes extremely topical and begins to haunt the minds of the most enlightened scientists: do we create our reality?

The study of the soul becomes part of physics.

Until the last century, every discourse on the existence of the soul was confined to the theological or philosophical sphere. Now, surprisingly, the question has become part of the agenda of theoretical physics, firmly placed at the top of the list of unsolved but worthy of in-depth topics.

In particular, physicist Henry Pierce Stapp is conducting studies declaimed with some emphasis on the web through titles similar to this one: "Henry Stapp wants to prove the existence of the soul" . In truth, Stapp does not want so much to demonstrate the existence of the soul, but rather that it, understood as the consciousness of the living being, cannot be ignored and must become an active and influential variable in the equations and formulas aimed at describing the functioning of the universe.

Stapp was born in 1928 in Cleveland, Ohio, USA. He earned his PhD in particle physics under the supervision of Nobel laureates Emilio Segrè and Owen Chamberlain at the University of California, Berkeley. He is known for his works in quantum mechanics, in particular for the development of the axiomatic S-matrix theory; he has published, among others, many articles relating to the non-local aspects of quantum mechanics and to Bell's theorem.

In 1958 he was invited by Wolfgang Pauli to the ETH Zurich to work with him personally on basic problems in quantum mechanics. Since 1969 he has been a direct collaborator of Werner Heisenberg at the Max Planck Institute in Munich, and since 1976 he has worked with John Wheeler. It is an academic career always carried out at the highest levels..

Henry Stapp. Quantum theory and free will.

In 2017 Stapp published the book "Quantum Theory and Free Will. How Mental Intentions Translate into Bodily Actions".

This book explains, in simple but precise terms, how orthodox quantum mechanics works and shows how this theory, interpreted realistically, assigns an important role to our conscious and free choices. Stapp starts from the principle that traditional biology and neuroscience still support essentially failed classical precepts, according to which mental intentions do not affect the behavior of matter. He shows how quantum mechanics provides a rational basis for a better understanding of the connections between consciousness and behavior, also allowing an explanation of certain phenomena currently considered "paranormal". These ideas have important implications for understanding ourselves and our mental processes, and therefore also for the significance of our lives.

In the book Stapp tackles one of the thorniest issues ever, that of free will. It is a concept that the "Encyclopedia Treccani" defines as follows:

> Free will. Expression used to indicate the freedom of man, whose acts are not determined by superior forces (of a supernatural or natural type), but derive from his autonomous choices.
>
> The expression was born on the ground of Christian theological discussions, in relation to the reconciliation between omnipotence and divine omniscience and human freedom. To support the thesis of free will with particular warmth were above all the protagonists of Humanism and the Renaissance.

From a religious point of view it is very difficult to reconcile free will with the idea of an omniscient God: if God already knows the future and therefore knows how man will act, can man consider himself free to make choices?

Let's put religion aside for a moment, since it is a personal and non-objective concept and that even believers of different religions often disagree about it.

Science, as we know, proceeds with its methods that we can define objectives: it carries out experiments, from which it draws concrete conclusions. If the experimental results conclusively confirm the hypothesis, this becomes a physical law, otherwise the judgment can be denied or suspended.

In the case of free will, no one has ever been able to claim that it is a valid law. In fact, it has never been possible to demonstrate that, by examining a certain situation several times, it always gives the same result..

Starting from this scientific basis we can formulate a broader reasoning, linked to the certainty that man is made of atoms and particles.

We know that atoms are in constant turmoil, and we know that, as part of this movement, they tend to aggregate with other atoms. But one thing we don't know: why do atoms attract and aggregate? According to what thrust three atoms (two of hydrogen and one of oxygen) join together to form a molecule of water? Is there a "conscience" that leads them to carry out this extremely useful form of aggregation, rather than the many other possible ones?

In fact, the atom shouldn't have a consciousness, so it shouldn't be able to "consciously" decide what kind of aggregation to make. Moreover, not even the water molecule can be defined as conscious. But the atoms can also aggregate in a more complex way, for example in order to form a biologically "alive" being, that is, surely endowed with consciousness, or at least instinct. Miraculously, from the aggregation of billions of atoms without consciousness, beings endowed with consciousness are born.

Because? It is a real puzzle. What can be the solution? Either the atoms, contrary to appearances, possess a consciousness, or they are guided by a consciousness external to them: in any case, in the aggregative process a pilot consciousness must exist.

From a religious point of view it could be said that everything is foreordained by God; the conscience from which all consciences and all conscious behaviors come is that of God.

However, science does not surrender to these metaphysical reasonings and continues its work to discover what are the mechanisms that govern the aggregative functioning of matter.

Understanding how and why atoms aggregate would mean discovering how the consciences of living beings are created, first of all that of man, and it would be concluded that the formation of consciousness takes place on the basis of a clearly definable process, if possible transferable in a formula. If so, the formation of consciousness would be part of a predictable and replicable process. This would establish the existence of "standard" consciences and it would be clear once and for all that free will does not exist.

Quantum physics develops a totally different view, which Stapp investigates in his book. We cannot observe a particle by establishing its position and velocity, or rather, we can only do so in a probabilistic way. By repeating the measurement several times with the same data, we find the particle in several different places. Not only that: observing the particle, we collapse the wave function that contains it, so we decide where it should be found.

In fact, the particle is suspended in a nebulous condition until, simply by observing it, that is, looking at it, we decide ourselves where to place it.

This has immeasurable ethical and philosophical consequences. If we decide, by observing them, where to place the particles, that is, if it is up to us to choose the characteristics of matter, then it means that free will exists. We are no longer machines dominated by matter, but beings capable of dominating matter with our will.

In one of his many interviews, Henry Stapp expresses himself on his quantum theory of consciousness:

> It is important for us, as individuals and collectively, to correct the currently widespread notion that science proves that we are mechanical automata. This idea is based on classical mechanics, now empirically invalidated. Its successor, quantum mechanics, represents us as psycho-physical beings whose conscious intentions are not determined by the physical aspects of Nature, but who nevertheless causally influence (they are the cause, ed.) The course of physical events.
>
> The question that immediately arises is this: "What is the connection between our conscious experiences and the physically described world?".
>
> The classical physical theories prevalent in science during the eighteenth and nineteenth centuries stated that any physical property is predetermined only by previous physical properties, with no added "input" from consciousness. This statement, if true, would imply that our lives are meaningless; it would reduce each of us to a substantially mindless mechanical automaton, unable to influence the course of future physical events either for our own benefit or for the benefit of anyone else.
>
> Today it is known that classical physical techniques are fundamentally false. They have been replaced by quantum mechanics. In this theory, our conscious "free

choices" play an essential dynamic role in determining the future. These choices are called "free" because they are not bound by any other characteristic of the theory, statistical or otherwise. "

The energy of the Spirit.

Our ability to observe elementary particles determines the reality of what we observe and therefore testifies to our free will. But then, we can ask ourselves: who is it that observes our particles to determine that they must exactly form our body, and therefore our consciousness? We, by whom are we observed?

Is there an observer who determines us? Or: is there a game in which we are observers and observed, creatures and creators? Who is driving this game? Or maybe the game comes from an "opening" that sees players and referees from the very beginning?

Maybe reality is a gigantic Ouroboros, the mythical dragon that bites its own tail, forming a circle with no beginning or end?

Apparently immobile, but in eternal movement, the ouroboros represents the power that devours and regenerates itself, the universal energy that is constantly consumed and renewed. It therefore symbolizes the unity and wholeness of the whole, infinity, eternity, cyclical time, eternal return, immortality and perfection.

It is curious to note how this symbol is present in a surprising multiplicity of mystical and philosophical traditions of every time and place.

The fact that all matter takes shape from observation contains an imperative necessity: we need to be observed.

George Berkeley's motto is illuminating again: Esse est percipi (To be means to be perceived).

From this perspective, the words of the psalmist also become understandable, who turns to God thanking him for being looked at, indeed, for not being able to escape his gaze. Not only that: recognizing that God looked at him ever since he was formed in his mother's womb, he admits that his gaze made him perfect "like a prodigy".

Probably, David Bohm as a child, as the son of a rabbi's assistant, would have prayed many times with the words of Psalm 138, especially since the author, King David, bore his name:

> "Lord, you scrutinize me and know me,
> you know when I sit and when I get up.
> Penetrate my thoughts from afar,
> you scrutinize me when I walk and when I rest.
> Where to go away from your spirit,
> where to escape from your presence?
> Still shapeless, your eyes saw me. "

In the last years of his life Bohm was fascinated by the theories of the philosopher Jiddu Krishnamurti, born in India in 1895. He established a cultural and spiritual partnership with him, so much so that they wrote together the book Where time ends.

In the course of this friendship Bohm came very close to Eastern mysticism and transformed his youthful desire to build a more just society (which had been spoiled by Soviet materialism) into an aspiration tending towards the same goal but through the path of spiritual growth.

Probably in the context of this cultural and spiritual partnership he elaborated the thought expressed in this quotation taken from the book "Bohm. The physics of infinity "by Massimo Teodorani:

> “The field of the finite is all that we can see, hear, touch, remember and describe. This field is basically what is manifest or tangible. The essential quality of infinity, of contrast, is its elusiveness, its intangibility. This quality is transmitted in the word Spirit, whose original meaning is "wind, breath".
>
> This suggests the existence of an invisible but pervasive energy, to which the manifest world of the finite

> responds. This energy, or Spirit, pervades all living beings, and without it every organism would break down into its constituent elements. What is really alive in the living being is this energy of the Spirit, and it was never born and never will die. "

Bohm was referring to the spirit understood as the Soul of the world, a super-implicated order underlying and vivifying of every subsequent implication and explanation, placed in the level of supreme freedom and, as such, free from any philosophical or dogmatic connotation.

Influenced, but not conditioned by the biblical imprint he received in his youth, Bohm would have imagined the spirit as a symbol of great freedom, as described in Genesis: "The earth was formless and deserted and darkness covered the abyss and the spirit of God hovered over the waters ".

As demonstrated in his life, Bohm considered freedom a great good and aspired to conquer it in its absoluteness in order to be able to share it. He died of a heart attack on October 27, 1992, in London, leaving a great void in the scientific community and a huge legacy of ideas waiting to be developed both scientifically and philosophically.

Bibliography.

Amir Dan Aczel, Entanglement. Il più grande mistero della fisica. Raffaello Cortina Editore.

Barbour Julian, La fine del tempo. Einaudi.

Barrow John David, Da zero a infinito. La grande storia del nulla. Oscar Mondadori.

Barrow John David, I numeri dell'universo, Oscar Mondadori

Barrow John David, Perché il mondo è matematico? Laterza

Barrow John David, Tipler Frank, Il principio antropico. Adelphi.

Beitman Bernard, I messaggi delle coincidenze. Sperling & Kupfer.

Cambray Joseph, Sincronicità. Natura e psiche in un universo interconnesso. Edizioni Fattore Umano

Cantalupi Tiziano, Santarcangelo Donato, Psiche e realtà. Tecniche Nuove.

Capra Fritjof, Il Tao della fisica. Adelphi.

Cederquist John, Le coincidenze non esistono. Rizzoli.

Cesati Cassin Marco, Non siamo qui per caso. Il potere delle coincidenze. Sperling & Kupfer.

Chandrasekhar Subrahmanyan, Verità e bellezza. Le ragioni dell'estetica nella scienza. Garzanti

Chinnici Giorgio, Guarda caso. I meccanismi segreti del mondo quantistico, Hoepli

Chopra Deepak, Le coincidenze. Sperling & Kupfer.

Ford Kenneth, Il mondo dei quanti. La fisica quantistica per tutti. Bollati Boringhieri.
Gamow George, Le avventure di mr. Tompkins. Edizioni Dedalo
Gamow George, Nuovo Mondo di mr. Tompkins. Zanichelli
Goswami Arneb, Guida quantica all'illuminazione. Edizioni Mediterranee
Greene Brian, La trama del cosmo. Spazio, tempo, realtà. Einaudi.
Greene Brian, La realtà nascosta – universi paralleli e leggi profonde del cosmo. Einaudi
Greene Brian, L' universo elegante. Superstringhe, dimensioni nascoste e la ricerca della teoria ultima. Einaudi.
Hawking Stephen, L'universo in un guscio di noce. Oscar Saggi Mondadori.
Hawking Stephen, La teoria del tutto. Origine e destino dell'universo. Mondadori. +
Hawking Stephen, La grande storia del tempo. Mondadori
Hawking Stephen, Dal Big Bang ai buchi neri. Breve storia dell'universo. Mondadori
Hawking Stephen, Mlodinov Leonard, Il grande disegno. Mondadori.
Heckler, Richard, Coincidenze. Lyra Libri.
Hopke Robert, Nulla succede per caso. Oscar Mondadori
Joseph Frank, Il potere delle coincidenze. Il punto d'incontro.
Jung Carl, L'analisi dei sogni. Gli archetipi dell'inconscio. La sincronicità. Bollati Boringhieri.
Jung Carl, Ricordi, sogni, riflessioni. Rizzoli

Kane Gordon, Il giardino delle particelle elementari. Longanesi
Kumar Manjit, Quantum. Da Einstein a Bohr, la teoria dei quanti, una nuova idea della realtà. Mondadori.
Kung Hans, Cristianesimo e religiosità cinese. Arnoldo Mondadori editore.
Lederman Leon, Hill Christopher, Fisica quantistica per poeti, Bollati Boringhieri.
Licata Ignazio, Osservando la Sfinge. Di Renzo Editore.
Motterlini Matteo, Trappole mentali. Rizzoli
Peat David, Sincronicità. Un connubio tra materia e psiche. Edizioni Magi
Popper Karl, L'Io e il suo cervello. Armando Editore.
Radin Dean, Menti interconnesse. I fenomeni psichici spiegati dalla fisica quantistica. Edizioni Mediterranee.
Rhine Louisa, Psicocinesi. La mente domina la materia. Golem Libri.
Schumacher Ernst, Una guida per i perplessi, Edizioni Piano B.
Sheldrake Rupert, Le illusioni della scienza. Urra.
Sheldrake Rupert, La mente estesa. Urra.
Smith Michael, Jung e lo sciamanesimo. Edizioni Amrita
Sparzani e Panepucci. (curatori)**,** Jung e Pauli. Il carteggio originale: l'incontro tra Psiche e Materia. Moretti & Vitali.
Stapp Henry, Quantum Theory and Free Will. Springer International Publishing.
Tagliagambe Silvano e Malinconico Angelo, Pauli e Jung. Un confronto su materia e psiche. Raffaello Cortina Editore.
Tagliagambe Silvano e Malinconico Angelo, Jung e il libro rosso: il Sé e il sacrificio dell'Io. Moretti & Vitali.

Talbot Michael, Tutto è uno. Feltrinelli
Teodorani Massimo, Bohm. La fisica dell'infinito. Macro Edizioni
Teodorani Massimo, La mente creatrice. Dall'universo fisico alla vita intelligente. Macro Edizioni.
Teodorani Massimo, Entanglement. L'intreccio nel mondo quantistico: dalle particelle alla coscienza. Macro Edizioni.
Teodorani Massimo, Sincronicità. Il legame tra fisica e psiche. Da Pauli a Jung a Chopra. Macro Edizioni.
Teodorani Massimo, L'atomo e le particelle elementari. Macro Edizioni
Tipler Frank, The physics of immortality. Doubleday.
White John, L'incontro tra scienza e spirito. Pratiche Editrice.
Widmann Claudio, Sincronicità e coincidenze significative. Edizioni Magi.
Widmann Claudio, Introduzione alla sincronicità. Edizioni Magi.

Finished printing on May 15, 2022

George Anderson is the pseudonym of Bruno Del Medico, blogger, writer, editor, specialized in the dissemination of issues related to social current events and the new frontiers of science. He is the author of many books on the recent Covid19 pandemic and of a series dedicated to quantum physics and metaphysics..